Intelligent Systems, Control and Automation: Science and Engineering

Volume 79

More information about this series at http://www.springer.com/series/6259

Spyros G. Tzafestas

Roboethics

A Navigating Overview

Springer

Spyros G. Tzafestas
School of Electrical and Computer
Engineering
National Technical University of Athens
Athens
Greece

ISSN 2213-8986 ISSN 2213-8994 (electronic)
Intelligent Systems, Control and Automation: Science and Engineering
ISBN 978-3-319-21713-0 ISBN 978-3-319-21714-7 (eBook)
DOI 10.1007/978-3-319-21714-7

Library of Congress Control Number: 2015945135

Springer Cham Heidelberg New York Dordrecht London

Printed on acid-free paper

Springer International Publishing AG Switzerland is part of Springer Science+Business Media
(www.springer.com)

To my loving grandchildren Philippos,
Myrto and Spyros

Preface

Morals are based on the knowledge of universal ideas, and so they have a universal character.
Plato

Relativity applies to physics, not to ethics.
Albert Einstein

The aim of this book is to provide a navigating introductory overview of the fundamental concepts, principles and problems in the field of *roboethics* (robot ethics). Roboethics is a branch of applied ethics that attempts to illuminate how the ethical principles can be applied to address the delicate and critical ethical questions arising in using robots in our society. Ethics has its origin in ancient philosophy that has put the analytic foundations on determining what is *right* and *wrong*. The Greek philosophers advocated models of life with the *human* as their central value ("*valeur par excellence*").

Robotics has been developed along these lines, i.e., robotics in the *service of mankind.* Robotics is directly connected to human life; medical robotics, assistive robotics, service-socialized robotics and military robotics all have strong impact on human life and pose major ethical problems for our society. Roboethics is receiving increasing attention within the fields of *techno-ethics* and *machine ethics*, and a rich literature is available that covers the entire spectrum of issues from theory to practice.

The depth and width of the presentation in this book is sufficient for the reader to understand the ethical concerns of designers and users of intelligent and autonomous robots, and the ways conflicts and dilemmas might be resolved. The book is of a tutorial nature, convenient for novices in the field, and includes some conceptual non-technical material of artificial/machine intelligence, the robot world with emphasis on the types and applications of robots, and mental robots that possess, besides cognition, intelligence and autonomy capabilities, consciousness and conscience features.

The book can be used both as a supplement in robotics courses and as a general information source. Those who are planning to study roboethics in-depth will find this book a convenient consolidated start.

I am deeply indebted to the Institute of Communication and Computer Systems (ICCS) of the National Technical University of Athens (NTUA) for supporting the project of this book, and to all colleagues for granting their permission to include in the book the requested pictures.

February 2015 Spyros G. Tzafestas

Contents

Chapter 1
Introductory Concepts and Outline of the Book

The best way of life is where justice and equity prevail.
Solon

The best way to teach morality is to make it a habit for the children.
Aristotle

1.1 Introduction

Roboethics is a new field of robotics which is concerned with both the positive and negative implications of robots to society. The term *roboethics* for "*robot ethics*" was coined by *Verrugio* [1]. Roboethics is the ethics that aims at inspiring the moral design, development and use of robots, especially intelligent/autonomous robots. The fundamental issues addressed by roboethics are: the dual-use of robots (robots can be used or misused), the anthropomorphization of robots, the humanization of human-robot symbiosis, the reduction of socio-technological gap, and the effect of robotics on the fair distribution of wealth and power [2]. According to the Encyclopaedia Britannica: "Robot is any automatically operated machine that replaces human effort, though it may not resemble human beings in appearance or perform functions in a human like manner". In his effort to find a connection between humans and robots *Gill* [3] concludes that: "Mechanically, human beings may be thought as direct-drive robots where many muscles play a role of direct drive motors. However, contradictory to science fiction, humans are much superior to robots in the structural point of view because the densities of muscles and bones of humans are an order lower than steel or copper, which are the major structural materials for robots and electrical motors".

The purpose of this chapter is:

- To provide a preliminary discussion of the concepts of roboethics and robot morality levels.
- To give a short literature review of roboethics.
- To explain the scope and provide an outline of the book.

S.G. Tzafestas, *Roboethics*, Intelligent Systems, Control and Automation: Science and Engineering 79, DOI 10.1007/978-3-319-21714-7_1

1.2 Roboethics and Levels of Robomorality

Nowadays there is a very rich literature on roboethics covering the whole range of issues from theoretical to practical roboethics for the design and use of modern robots. Roboethics belongs to techno-ethics which deal with the ethics of technology in general, and to machine ethics which extends computer ethics so as to address the ethical questions in designing and using intelligent machines [4, 5]. Specifically *roboethics* aims at developing scientific, technical and social ethical systems and norms related with the creation and employment of robots in society. Today, in advanced research in computer science and robotics, the effort is to design *autonomy* which is interpreted as the ability required for the machines and robots to carry out autonomously intellectual human-like tasks. Of course, autonomy in this context should be properly defined since it might be misleading. In general, "autonomy is the capacity to be one's own person to live one's life according to reasons and motives taken as one's own and not the product of external forces" [6].

Autonomy in machines and robots should be used in a narrower sense than humans (i.e., metaphorically). Specifically, machine/robot autonomy cannot be defined absolutely, but only relatively to the goals and tasks required. Of course we may frequently have the case in which the results of the operations of a machine/robot are not known in advance by human designers and operators. But this does not mean that the machine/robot is a (fully) autonomous and independent agent that decides what to do by its own. Actually, machines and robots can be regarded as partially autonomous agents, in which case we may have several levels of "*autonomy*" [7]. The same is true for the issues of "*ethics*", where we have several levels of "*morality*" as described in Sect. 5.4, namely [8]:

- Operational morality
- Functional morality
- Full morality

Actually, it would be very difficult (if not impossible) to describe ethics with sufficient precision for programming and fully embedding it to a robot. But, the more autonomy a robot is provided and allowed to have, the more morality (ethical sensitivity) is required by the robot.

In general, ethics within robotics research must have as central concern to warn against the negative implications from designing, implementing and employing robots (especially autonomous robots). This means that, in actual life, roboethics should provide the moral tools that promote and encourage the society and individuals to keep preventing misuse of the technological achievements in robotics against the human kind. Legislation should provide *efficient and just* legal tools for discouraging and preventing such misuse, and assigning liability in case of harm due to robot misuse and human malpractice.

1.3 Literature Review

We start with a brief outline of the seminal Special Issue "*Ethics in Robotics*" of the International Review of Information Ethics (Vol. 6, December 2006). This issue, edited by *R. Capurro*, contains thirteen contributions, offered by eminent researchers in the field of roboethics, that cover a wide set of fundamental issues.

G. Veruggio and F. Operto present the so-called "Robotics Roadmap" [9], which is the result of a cross-cultural interdisciplinary discussion among scientists that aim at monitoring the effects of current robotic technologies to society.

P.M. Asaro deals with the question "what we want from a robot ethic", and argues that the best approach to roboethics is to take into account the ethics built into robots, the ethics of those designing and using robots, and the ethics of robot use.

A.S. Duff is concerned with justice in the information age following the neo-Pawisian approach for the development of a normative theory for information society. Aspects that are suggested to be considered include political philosophy, social and technological issues, the rights prior to good, social well-being and political stability.

J.P. Sullins investigates the conditions for a robot to be a moral agent, and argues that the questions which must be addressed for the evaluation of robot's moral status are: (i) Is robot significantly autonomous? (ii) Is the robot behaviour intentional?, and (iii) Is the robot in a position of responsibility?

B.R. Duffy explores the fundamental differences of humans and robots in the context of social robotics, and discusses the issue of understanding how to address them.

B. Becker is concerned with the construction of embodied conversational agents (robots and avatars) for human-computer interface development. She argues that this construction aims to provide new insights in the cognition and communication based on the creation of intelligent artifacts, and on the "idea" that such a mechanical human-like dialog will be beneficial for the human-robot interaction. The actual plausibility of this is put as an issue of discussion.

D. Marino and G. Tamburini are concerned with the moral responsibility and liability assignment problems in the light of epistemic limitations on prediction and explanation of robot behaviour that results by learning from experience. They argue that roboticists cannot be freed from all responsibility on the sole ground that they do not have full control over the causal chains implied by the actions of their robots. They rather suggest to use legal principles so as to fill the responsibility "gap" that some authors accept to exist between human and robot responsibility (i.e., that the greater the robot autonomy the less responsible the human).

C.K.M. Grutzen explores the vision of future and daily life in "ambient intelligence" (**AmI**). The assumption is made that intelligent technology should disappear into our environment in order to bring humans an easy and entertaining life. He argues that to investigate whether humans are in danger to become just objects of artificial intelligence, the relation between mental, physical, methodical invisibility and visibility of **AmI** should be examined.

S. Krebs investigates the influence of "mangas" (comics) and "animes" on the social perception and cultural understanding of robots in Japan. This includes the interaction between population culture and Japanese robotics. The *Astro* boy comics is used in order to examine the ethical conflicts between robots and humans that occur in Japanese mangas.

M. Kraehling investigates how Sony's robodog *Aibo* challenges the interpretation of other life forms by humans and how concepts of friendship are held. He argues that ethical issues about human perceptions of dogs in the era of doglike robots must be investigated, and that Aibo itself does not belong to a predefined category. It actually belongs somewhere in an intermediate space, i.e. it does not live in a really mechanistic world not in the animals world.

N. Kitano investigates the motivation for the popularity of socialized robots in Japan. The effects of them on human relations, and the customs and psychology of the Japanese are considered. First, the Japanese social context is described to illustrate the term "*Rinri*" (ethics, social responsibility). Then the meaning of Japanese "*Animism*" is explained to understand why Rinri is to be considered as an incitement of Japanese social robots.

M.A. Pérez Alvarez is concerned with the transformation of the educational experiences into the class rooms which are necessary for the development of intellectual abilities of children and teenagers. It is argued that the process of "arming and programming" *LEGO-type Mind Storms* enhances the young people's live experiences that favour the development of intelligence, and so it provides a useful education process in the class room.

D. Söffler and J. Weber discuss the question whether an autonomous robot, designed to communicate and make decisions in a human-like way is still a machine. The concepts, ideas and values on which such robots are based are discussed. The way they relate to everyday life and how far the social demands drive the development of such machines are examined. The question "whether the human-robot relationship changes" was investigated via an e-mail dialogue on ethical and socio-political aspects, especially on private life.

Some of the concepts and results derived in the above works will be discussed further in the book. We now continue with reviewing a few other important works on roboethics.

In [8], *Wallach and Allen* consider the question "can a machine be a genuine cause of harm? They argue and conclude that the answer is affirmative. They predict that within a number of years "there will be a catastrophic incident brought about by a computer system making a decision independent of human oversight". They examine deeply the need of machine morality, where we are at present, and conclude that this could be achieved.

In [10], *Capurro* recapitulates a number of his researches regarding the epistemological, ontological, and psychoanalytic implications of the relation between humans and robots, and about the ethical issues of human-machine interaction. He argues that the human-robot relation can be regarded as an envy relation in which humans either envy robots for what they are or they envy other humans for having robots that they do not have. Regarding the ethics of man-machine interaction the following questions are addressed:

(i) How do we live in a technological environment?
(ii) What is the impact of robots on society?
(iii) How do we (as users) handle robots? What methods and means are used today to model the interface between man and machine?

In [11], *Lin, Abney and Bekey* bring together prominent researchers and professionals from science and humanities to explore questions like: (i) Should robots be programmed to follow a code of ethics, if this is possible? (ii) How might society and ethics change with robots? (iii) Are there risks in developing emotional bonds with robot? (iv) Should robots, whether biological, computational hybrids, or pure machines be given rights or moral consideration? Ethics seems to be slow to follow the technological progress, and therefore the opinions of the contributors to the book are very helpful for the development of a "roboethics".

In [12], *Fedaghi* proposes a classification scheme of ethical categories to simplify the process by which a robot may determine which action is most ethical in complicated situations. As an application of this scheme Asimov's robot laws are decomposed and rephrased to support logical reasoning. Such an approach is in-line with the so-called procedural ethics.

In [13], Powers proposes a rule-based robot ethics system based on the assumption that the Kantian ideological/deontological ethical code can be reduced to a set of basic rules from which the robot can produce new ethical rules suitable to face new circumstances. Kantian ethics states that moral agents are both rational and free. But, as argued by many authors, embedding ethical rules in a robot agent naturally limits its freedom of thought and reasoning.

In [14], *Shibata, Yoshida and Yamato*, discuss the issue of using robotic pets in therapy of the elderly via some level of companionship. They discuss as a good representative example for this application the seal robot *Paro*, which has also been extended for use as part of therapeutic sessions in pediatric and elderly institutions world-wide [15].

In [16], *Arkin* summarizes the ethical issues faced in three realities, namely autonomous robots capable of lethal action, entertainment robots, and unemployment due to robotics. He argues that in the first reality (lethality by autonomous robots) the international laws of war and rules of engagement must be strictly followed by the robots. To assure this, the *Just War* theory should be understood, and methods should be developed and delineated for combatant/non-combatant discrimination. For the second area (personal robotics) he argues that a deep understanding of both the robot capabilities and human psychology is needed, in order to explore whether the roboticists' goal to induce pleasant psychological states can be achieved. The third area concerning robotics and unemployment is of social concern since the time where industrial robots were put in action (in shipyards and other manufacturing environments). It is argued that a clash between utilitarian and deontological morality approaches should be followed in order to deal with both the industrial/manufacturers' concerns and the rights of the individual workers.

In [17] *Huttunen* and colleagues discuss the legal perspective of responsible robots. Their work does not focus on ethical considerations, but on the legal liability risks related to inherently error-prone intelligent machines. A solution combining legal and economical aspects is proposed. To overcome the difficulties of creating perfectly functioning machines and the cognitive element inherent in intelligent machines and human-machine interaction a new kind of legal approach is developed (a financial instrument liberating the machine). In this way a machine can become an ultimate machine by emancipating itself from its manufacturer, owner, and operator.

In [18], *Murphy and Woods* have rephrased Asimov's laws (which they view as robot-centric) such as to remind robotics researchers and practitioners of their professional responsibilities. Asimov's laws placed in the scheme of morality levels mentioned in Sect. 2.1 [8], regard the robots to have functional morality, i.e., robots have sufficient agency and cognition to make moral decisions. The alternative laws proposed are more feasible to implement than Asimov's laws with current technology but they also raise new questions for investigation (see Chap. 5, Sect. 5.3).

In [19], *Decker* addresses the question of whether humans can be replaced in specific contexts of action by robots, on the basis of an interdisciplinary technology assessment. Adopting a "means-end" approach the following types of replacement are investigated: (i) technical replacement, (ii) economic replacement, (iii) legal replacement, and (iv) ethical replacement. The robots considered in the study are assumed to have advanced modern learning capabilities. Regarding autonomy levels the following levels are employed: (i) first-level (technical) autonomy, (ii) second-level (personal) autonomy, and (iii) third-level (ideal) autonomy. The conclusion of this study is that in a Kantian ethical perspective robot learning should be assigned as the responsibility of the robot's owner.

In [20], *Lichocki, Kahn Jr. and Billard*, provide a comprehensive survey of a number of basic ethical issues pertaining to robotics. The question of whom or what is responsible when robots cause a harm is first discussed. Then, the ethical aspects emerging in lethal robots created to act in the battlefields, and in service robots are discussed. In all cases, investigators agree that they want robots which contribute to a better world. The disagreements are on how this could be achieved. Some people want to embed (and actually embed) moral rules in the robot controller, while others argue against this, asserting that robots themselves cannot become moral agents. Others explore the use of robots for helping children with autism or for assisting elderly. The questions addressed in the studies reviewed range from philosophical ones, and extend to psychological and legislation questions.

Two comprehensive books on lethal robot ethics are [21, 22], and three important contributed book chapters on related topics for autonomous learning and android systems are provided in [23–25]. Books on the general field of computer and machine ethics include [26–28]. Three recent books addressing machine, information, and robot ethics questions are [29–31].

In [26], an in-depth exploration and analysis of the ethical implications of widespread use of computer technology is provided bringing together philosophy, law, and technology. In [27] the wider field of computer/machine morality is investigated including key topics such as privacy, software protection, artificial intelligence, workspace issues, and virtual reality. In [28] an anthology of 31 well-selected contributions is provided written by competent researchers in the field of ethics for machines. In [29] the ethical perspective considered, is the one that humans have when interacting with robots, including health care and warfare robotic applications and the moral aspects of human-robot cooperation. Finally, in [30] the question of machine moral agency is addressed, i.e., whether a machine might have moral responsibility for decisions and actions. The question whether a machine might be a moral patient legitimate of moral consideration is also investigated.

1.4 Outline of the Book

We have seen that roboethics is concerned with the examination and analysis of the ethical issues associated with the design and use of robots that possess a certain level of autonomy. This autonomy is achieved by employing robot control and artificial intelligence techniques. Therefore *roboethics* (**RE**) is based on three field components, namely: *ethics* (**E**), *robotics* (**R**), and *artificial intelligence* (**AI**) as shown in Fig. 1.1.

In practice, roboethics is applied to the following subfields that cover the activities and applications of modern society as shown in Fig. 1.2.

- Medical roboethics
- Assistive roboethics
- Service/socialized roboethics
- War roboethics

Based on the above elements (shown in Figs. 1.1 and 1.2) the book involves 12 chapters including the present chapter. Chapters 2–5 provide background material and deal with the fields of ethics, artificial intelligence, robotics, and roboethics in

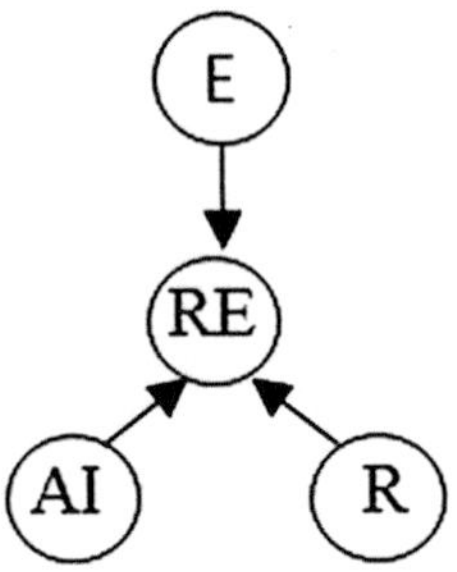

Fig. 1.1 Robotethics (*RE*) and its three contributing fields: ethics (*E*), robotics (*R*), and artificial intelligence (*AI*)

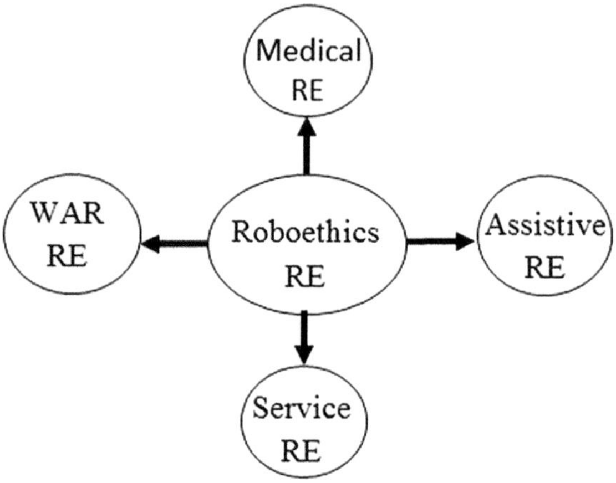

Fig. 1.2 The four principal areas of roboethics

general. Chapters 6–9 provide an exposition of medical, assistive, socialized, and war roboethics. Chapter 10 provides an overview of roboethics as conceived in Japan, and some intercultural issues concerning roboethics and infoethics. Chapter 11 discusses three further topics of roboethics, namely: autonomous (self-driving) cars, cyborgs, and privacy roboethics. Finally, Chap. 12 provides a short review of mental robots and their abilities.

The topic of the social and ethical implications of industrial robotics is fully discussed in most classical textbooks on robotics (e.g., [32, 33]). The principal critical issues of industrial robotics that concern the human individuals and society are:

- **Training and education** More well-educated robotics experts and operators are still needed. Most persons in the society either don't trust robotics or over trust it. Both are not good. Therefore people should be informed in a realistic and reliable way about the potential capabilities and risks of advanced robotics.
- **Unemployment** This was the most important issue of discussion two to three decades ago, but presently is at an acceptable equilibrium level due to increasing generation of new jobs.
- **Quality of working conditions** Working conditions are improved when robotics is used for jobs of the so-called three Ds, namely "*Dirty, Dangerous, and Dull*". Productivity increases may also, in the long term, result in a shorter and more flexible scheduled work week for the benefit of the workers.

The ethical aspect of robotics scientists for the unemployment should be of continuous concern. Robotics and automation engineers have the ethical duty to exert as much as they can their influence for social support for those potentially made unemployed. As for the working condition's quality the ethical duty of engineers is to develop the most efficient safety systems for human protection in all environments of robotics use, especially when the robots are in direct physical contact with humans. The outline of the book is as follows.

Chapter 2, "Ethics: Fundamental Elements", presents the fundamental concepts and theories of ethics and applied ethics. The ethics branches, namely meta-ethics, normative ethics, and applied ethics are discussed. Specifically, the following ethics theories are considered: virtue theory (*Aristotle*), deontological theory (*Kant*), utilitarian theory (*Mill*), justice as fairness theory (*Rawls*), egoism theory, value-based theory, and case-based (casuistry) theory. The chapter ends with a discussion of professional ethics.

Chapter 3, "Artificial Intelligence", provides a sketch of the central concepts and issues of artificial intelligence, namely the differences between human and artificial intelligence, the Turing intelligence test, the methods of applied artificial intelligence, and the human-computer interaction/interfaces topic.

Chapter 4, "The World of robots", provides a tour to the world of robots, explaining the types of robots by kinematic structure and locomotion, and the artificial intelligence tools that give intelligence capabilities to robots. Then, the chapter discusses the robot applications (medicine, society, space, military) for which the ethical issues are addressed in Chaps. 6–10.

Chapter 5, "Roboethics: A Branch of Applied Ethics", outlines a set of preliminary issues of roboethics, discusses the "top-down" (deontological, consequentialist) methodology and the "bottom-up" (learning-based) approach to roboethics. Then, the chapter presents the basic requirements for a smooth human-robot symbiosis, and discusses the issue of "robot rights".

Chapter 6, "Medical Roboethics", is concerned with the ethics of medical robotics, which is the most fundamental subfield of roboethics, with immediate positive effects on human life. The chapter starts with a short discussion of medical ethics (in general). Then, it provides an outline of the basic technological aspects and the particular ethical issues of robotic surgery.

Chapter 7, "Assistive Roboethics", starts with a discussion of a number of assistive devices, as a background, and then outlines the basic ethical principles and guidelines of assistive robotics including the ethical codes of two rehabilitation and assistive technology societies.

Chapter 8, "Socialized Roboethics", covers the class of service robots with emphasis on socialized robots. Specifically, it presents the various definitions of socialized robots, along with a number of examples. Then, it discusses the fundamental ethical issues of socialized robots, and reviews three case studies of children-robot and elderly-robot therapeutical/social interactions.

Chapter 9, "War Roboethics", is concerned with the ethics of using robots, especially autonomous lethal robots, in warfare. The chapter starts with basic material on the general ethical laws of war, and ends with an exposition of some arguments against autonomous war robots, as well as some counterarguments.

Chapter 10, "Japanese Roboethics, Intercultural, and Legislation Issues", presents the fundamental aspects of the traditional ethical perspective of robots in Japan, starting with a short overview of the indigenous Japanese culture and ethics. Then, a discussion of some intercultural issues of infoethics/roboethics is provided based on the shared norms and shared values approaches. The chapter ends with an outline of the legislation of robots in the West and East.

Chapter 11, "Additional Roboethics Issues", discusses the ethical concerns arising in the design and use of autonomous (self-driving, driverless) cars, and cyborgs (cybernetic organisms), as well as the "privacy" ethical issues of modern robots. Autonomous cars roboethics is analogous to surgical robots ethics and the ethical/legal responsibility in case of harm. The principal application of cyborg technology is in medicine (restorative and enhanced cyborgs). Regarding privacy, the chapter is concentrated to the new ways social robots, equipped with several sophisticated sensors, can implicate the privacy of their users.

Chapter 12, "Mental Robots", complements the material of Chaps. 3 and 4, providing a conceptual study of "mental robots" and their "brain-like" capabilities, namely: cognition, intelligence, autonomy, consciousness, and conscience/ethics, and discussing the features of the more specialized processes of learning and attention.

In overall, the book gives a spherical picture of the current status of the new field of roboethics covering the most fundamental concepts, questions, and issues. The field is in its first period of development and many more results are expected to be derived by the ongoing research activity in machine ethics that goes in parallel with the progress of artificial intelligence.

References

1. Veruggio G (2005) The birth of roboethics. In: Proceedings of IEEE international conference on robotics and automation (ICRA 2005): workshop on robo-ethics, Barcelona, pp 1–4
2. Veruggio G, Operto F, Roboethics: a bottom-up interdisciplinary discourse in the field of applied ethics in robotics. Int Rev Inf Ethics (IRIE), **6**(12):6.2–6.8
3. Gill LD (2005) Axiomatic design and fabrication of composite structures. Oxford University Press, Oxford
4. Allen C, Wallach W, Smit I (2006) Why machine ethics? IEEE Intell Syst 21(4):12–17
5. Hall J (2000) Ethics for machines. In: Anderson M, Leigh Anderson S (eds) Machine ethics. Cambridge University Press, Cambridge, pp 28–46
6. Christian J (2003) Autonomy in moral and political philosophy. In: Zalta EN (ed) The stanford encyclopedia of philosophy, Fall 2003 edn. http://plato.stanford.edu/archives/fall2003/entries/autonomy-moral/
7. Amigoni F, Schiaffonati V (2005) Machine ethics and human ethics: a critical view. AI and Robotics Lab., DEI, Politecnico Milano, Milano, Italy
8. Wallach W, Allen C (2009) Moral machines: teaching robots right from wrong. Oxford University Press, Oxford
9. Veruggio G (2006) The EURON roboethics roadmap. In: Proceedings of 6th IEEE-RAS international conference on humanoid robots, Genova, Italy, 4–6 Dec 2006, pp 612–617
10. Capurro R (2009) Ethics and robotics. In: Proceedings of workshop "L'uomo e la macchina", University of Pisa, Pisa, 17–18 May 2007. Also published in Capurro R, Nagenborg M (eds) Ethics and robotics. Akademische Verlagsgesellschaft, Heidelberg, pp 117–123
11. Lin P, Abney K, Bekey G (eds) (2012) Robot ethics: the ethical and social implications of robotics. MIT Press, Cambridge, MA
12. Al-Fedaghi SS (2008) Typification-based ethics for artificial agents. In: Proceedings of 2nd IEEE international conference on digital ecosystems and technologies (DEST), Phitsanulok, Thailand, 26–28 Feb 2008, pp 482–491

13. Powers TM (2006) Prospects for a Kantian machine. IEEE Intell Syst 21(4):46–51
14. Shibata T, Yoshida M, Yamato J (1997) Artificial emotional creature for human machine interaction. In: Proceedings of 1997 IEEE international conference on systems, man, and cybernetics, vol 3, pp 2269–2274
15. Wada K, Shibata T, Musha T, Kimura S (2008) Robot therapy for elders effected by dementia. IEEE Engineering in Medicine and Biology 27(4):53–60
16. R. Arkin, On the ethical quandaries of a practicing roboticist: A first-hand look, In : A Briggle, K. Waelbers and P. Brey (eds), Current Issues in Computing and Philosophy, Vol. 175, Frontiers in Artificial Intelligence and Applications, Ch.5, Amsterdam: I0S Press; 2008
17. Arkin R (2008) On the ethical quandaries of a practicing roboticist: a first-hand look. In: Briggle A, Waelbers K, Brey P (eds) Current issues in computing and philosophy, vol 175, Frontiers in artificial intelligence and applications, Ch. 5, IOS Press, Amsterdam
18. Hurttunen A, Kulovesi J, Brace W, Kantola V (2010) Liberating intelligent machines with financial instruments. Nord J Commer Law 2:2010
19. Decker M (2007) Can humans be replaced by autonomous robots? Ethical reflections in the framework of an interdisciplinary technology assessment. In: IEEE robotics and automation conference (ICRA-07), Italy, 10–14 Apr 2007
20. Lichocki P, Billard A, Kahn PH Jr (2011) The ethical landscape of robotics. IEEE Robot Autom Mag 18(1):39–50
21. Arkin RC (2009) Governing lethal behavior in autonomous robots. Chapman Hall/CRC, New York
22. Epstein RG (1996) The case of the killer robot: Stories about the professional, ethical and societal dimensions of computing. John Wiley, New York
23. Kahn A, Umar F (1995) The ethics of autonomous learning system. In: Ford KM, Glymour C, Hayes PJ (eds) Android epistemology. The MIT Press, Cambridge, MA
24. Nadeau JE (1995) Only android can be ethical. In: Ford KM, Glymour C, Hayes PJ (eds) Android epistemology. The MIT Press, Cambridge, MA
25. Minsky M (1995) Alienable rights. In: Ford K, Glymour C, Hayes PJ (eds) Android epistemology. The MIT Press, Cambridge, MA
26. Johnson DG (2009) Computer ethics. Pearson, London
27. Elgar SL (2002) Morality and machines: perspectives on computer ethics. Jones & Bartlett, Burlington, MA
28. Anderson M, Leigh Anderson S (eds) (2011) Machine ethics. Cambridge University Press, Cambridge, UK
29. Capurro R, Nagenborg M (2009) Ethics and robotics. IOS Press, Amsterdam
30. Gundel DJ (2012) The machine question: critical perspectives on AI, robots and ethics. The MIT Press, Cambridge, MA
31. Dekker M, Guttman M (eds) (2012) Robo-and-information ethics: some fundamentals. LIT Verlag, Muenster
32. Hunt VD (1983) Industrial robotics handbook. Industrial Press, Inc., New York
33. Groover MP, Weiss M, Nagel RW, Odrey NG (1986) Industrial robotics: technology, programming and applications. McGraw-Hill, New York

Chapter 2
Ethics: Fundamental Elements

Ethics is to know the difference between what you have the right to do and what is right to do.

Potter Stewart

With respect to social consequences I believe that every researcher has responsibility to assess, and try to inform others of the possible social consequences of the research products he is trying to create.

Herbert Simon

2.1 Introduction

Ethics deals with the study and justification of moral beliefs. It is a branch of philosophy which examines what is *right* and what is *wrong*. *Ethics* and *More* are regarded as identical concepts, but actually they are not. The term ethics is derived from the Greek word *ήθος* (ethos) meaning *moral character*. The term *morality* comes from the Latin word *mos* meaning *custom* or *manner*. Morals, from which the term morality is derived, are social rules or inhibitions from the society. In present times this is, in a way, reverted, i.e., ethics is the science, and morals refer to one's conduct or character. *Character* is an *inner-driven* view of what constitutes morality, whereas *conduct* is an *outer-driven* view. Philosophers regard ethics as moral philosophy and morals as societal beliefs. Thus it might happen that some society's morals are not ethical, because they represent merely the belief of the majority. However, there are philosophers who argue that ethics has a relativistic nature, in the sense that what is right is determined by what the majority believe [1–3]. For example, in ancient Greece Aristotle's view of ethics was that "ethical rules should always be seen in the light of traditions and the accepted opinions of the community".

Some psychologists such as Lawrence Kohlberg argue that moral behavior is derived by moral reasoning which is based on the principles and methods that one uses in his/her judgment. Other psychologists regard the ethical behavior as the

S.G. Tzafestas, *Roboethics*, Intelligent Systems, Control and Automation: Science and Engineering 79, DOI 10.1007/978-3-319-21714-7_2

humanistic psychology movement. For example, to determine what is right and wrong, one may start from "*self-actualization*" which is one's highest need and fulfils his/her potential. Still other psychologists have developed the *evolutionary psychology* which is based on the assumption that the ethical behavior can sometimes be seen as an evolutionary process. For example, altruism towards members of one's own family promotes his/her inclusive fitness.

The objective of this chapter is to present the fundamental concepts and issues of ethics in general. In particular, the chapter:

- Discusses the branches of analytic philosophy ethics.
- Investigates the principal theories of ethics.
- Discusses the issue of professional ethics and presents the codes of ethics for engineers, electrical and electronic engineers, and robotic engineers.

2.2 Ethics Branches

In analytic philosophy, ethics is distinguished in the following levels:

- Meta ethics
- Normative ethics
- Applied ethics

2.2.1 Meta Ethics

Meta ethics is one of the fundamental branches of philosophy which examines the nature of morality in general, and what justifies moral judgments. Three questions investigated by meta ethics are:

- Are ethical demands *true-apt* (i.e., capable of being true or not true) or are they, for example, emotional claims?
- If they are *true-apt*, are they ever true, and if so what is the nature of the facts they represent?
- If there are moral truths what makes them true, and they are absolutely true or always relative to some individual or society or culture?

If there are more truths, one way to find what makes them true is to use a *value system*, and here the question is if there is a value that can be discovered. The ancient Greek philosophers, e.g., *Socrates* and *Plato* would reply *yes* (they both believed that goodness exists absolutely), although they did not have the same view about what is good. The view that there are no ethical truths is known as "*moral anti-realism*". The modern empiricist *Humes* has the position that moral expressions are expressions of emotion or sentiment feeling. Actually, the value system of a society is created by great individuals (writers, poets, artists, leaders, etc.) or derived from some list of *moral absolutes*, e.g. religious moral code, whether explicit or not.

2.2.2 Normative Ethics

Normative ethics studies the issue of how we ought to live and act. A normative ethics theory of the good life investigates the requirements for a human to live well. A normative theory of right action attempts to find what it is for an action to be morally acceptable.

In other words normative ethics attempts to provide a system of principles, rules and procedures for determining what (morally speaking) a person should do and should not do. Normative ethics is distinguished from meta-ethics because it investigates standards for the rightness and wrongness of actions, whereas meta-ethics examines the meaning of moral language and the metaphysics of moral facts. Normative ethics is also different from "descriptive ethics" which is an empirical investigation of people's moral beliefs.

Norms are sentences (rules) that aim to affect an action, rather than conceptual abstractions which describe, explain, and express. Normative sentences include commands, permissions and prohibitions, while common abstract concepts include sincerity, justification, and honesty. Normative rules interpret "ought-to" kind of statements and assertions, as contrasted from sentences that give "is" type statements and assertions. A typical way to "normative ethics" is to describe "norms" as reasons to believe, and to feel.

Finally, a theory of *social justice* is an attempt to find how a society must be structured, and how the social goods of freedom and power should be distributed in a society.

2.2.3 Applied Ethics

Applied ethics is the branch of ethics which investigates the application of ethical theories in actual life. To this end, applied ethics attempts to illuminate the possibility of disagreement about the way theories and principles should be applied [4]. Specific areas of applied ethics are:

- Medical ethics
- Bioethics
- Public sector ethics
- Welfare ethics
- Business ethics
- Decision making ethics
- Legal ethics (justice)

- Media ethics
- Environmental ethics
- Manufacturing ethics
- Computer ethics
- Robot ethics
- Automation ethics

Strict deontological principles of 'Ten Commandments' type provides solutions to particular cases that are not globally acceptable. For example, in medical ethics a strict deontological approach would never allow the deception of a patient about his/her illness, whereas a utilitarian approach might permit lying to a patient if the outcome of the deception is good.

2.3 Ethics Theories

Key ethical theories are the following:

- Virtue theory (*Aristotle*)
- Deontological theory (*Kant*)
- Utilitarian theory (*Mill*)
- Justice as fairness theory (*Rawls*)
- Egoism theory
- Value-based theory
- Case-based theory

2.3.1 Virtue Theory

Aristotle's ethical theory is based on the concept of *virtue* which is defined to be a character a human being needs to flourish or live well. *Virtue* is coming from the Latin word *virtus* and the Greek *αρετή* (areti) meaning excellence of a person. A *virtuous agent* is one who has and applies the virtues (i.e., an agent that acts virtuously). Virtue theory tells that "an action is right if it is what a virtuous agent would do in the situation at hand" [5, 6]. Thus, virtue theory is actually concerned with building good personality (character) by creating traits and habits toward acting with justice, prudence, courage, temperance, compassion, wisdom, and fortitude. The character (model of practical reasoning) is built by answering the question "what habits should I develop?" In overall, the creation of personal identity is achieved by combining desires, reason and character habits. Aristotle's two principal virtues are "*σοφία*" (sophia) meaning *theoretical wisdom* and "*φρόνησις*" (phronesis) meaning *practical wisdom*. Plato's cardinal virtues are *wisdom, courage, temperance,* and *justice,* i.e., if one is wise, courageous, temperate and just, then right actions will follow.

2.3.2 Deontological Theory

Kant's ethical theory [7] gives emphasis to the principles upon which the actions are based rather than the actions' results. Therefore, to act rightly one must be motivated by proper universal deontological principles that treat everyone with respect (respect for persons theory). The term *deontology* is derived from the Greek word "*δεοντολογία*" (deontology) which is composed by two words "*δέον*" (deon = duty/obligation/right) and "*λόγος*" (logos = *study*). Thus deontology is the ethical theory based on duties, obligations and rights. When one is motivated by the right principles he/she overcomes the animal's instincts and acts ethically. The center of Kant's ethics is the concept of "*categorical imperative*". His model of practical reasoning is based on the answer to the question: "how do I determine what is rational?" Here, rationality means "do what reason requires" (i.e., without inconsistent or self-contradictory policies). Another approach to deontological theory is *Aquinas natural law* [8]. A further formulation of deontology is: "act such that you treat humanity, both in yourself and in that of another person, always as an end and never as a means". Persons, unlike things, ought never merely be used. They are ends in themselves.

The reason why Kant does not base ethics on consequences of actions but to duties is that, in spite of our best efforts, we cannot control the future. We are praised or blamed for actions within our control (which includes our will or intention) and not for our achievements. This does not mean that Kant did not care about the outcomes of actions. He is simply insisted that for a moral evaluation of our actions, consequences do not matter.

2.3.3 Utilitarian Theory

This theory, called *Mill's ethical theory,* belongs to the consequentialism ethical theories that are "*teleological*", which aim at some goal state and evaluate morality of actions toward the goal. More specifically, *utilitarianism* measures morality on the basis of the maximization of net expected utility for everyone affected by a decision or action. The fundamental principle of utilitarianism can be stated as [9]:

> Actions are moral to the extent that they are oriented towards promoting the best long-term interests (greatest good) for everyone concerned.

Of course, in many cases it is not clear what constitutes the "*greatest good*". Some utilitarians consider that what is intrinsically good is pleasure and happiness, while others say that other things are intrinsically good, namely beauty, knowledge and power.

According to *Mill* not all pleasures have equal worth. He defined the "good" in terms of *well-being* (pleasure or happiness), which is the Aristotelian *ευδαιμονία* (eudaimonia = happiness). He distinguished happiness not only quantitatively but also qualitatively between various forms of pleasure.

The utility principle tries to bridge the gap between *empirical* facts and *normative conclusions* using a pure cost/benefit analysis. Here, each one person should be counted as only one and no one person is allowed to be counted as more than one. Drawbacks (difficulties) of utilitarianism include the following:

- It is not always possible to determine who is affected by the outcome of an action.
- An outcome may not be the result of a unique action.
- The pleasures cannot easily be quantified using cost/benefit analysis.
- The greatest good for the greatest number is specified in "aggregate" way. Therefore this good may be obtained under conditions that are harmful to some ones.
- The process of determining what is right (or wrong) is a complex and time-consuming process.

2.3.4 *Justice as Fairness Theory*

This theory was developed by *John Rawls* (1921–2002). He combined the Kantian and utilitarian philosophies for the evaluation of social and political bodies. The *justice as fairness* theory is based on the following principle [10]:

> General primary goods-liberty and opportunity, income and wealth, and the bases of self-respect – are to be distributed equally, unless an unequal distribution of any or all of these goods is to the advantage of the least favored.

This principle involves two parts: the *liberty principle* (each human has an equal right to the widest basic liberty compatible with the liberty of others) and the *difference principle* (economic and social inequalities must be regulated such as they are reasonably expected to be to everyone's benefit attached to positions and offices to all). The Kantian liberty principle calls for universal basic respect for people as a minimum standard for all institutions. The difference principle suggests that all actions may be to the economic and social advantage of all, especially the least favored (like the utilitarian theory) with reasonable differences allowed.

2.3.5 *Egoism Theory*

Egoism theory is a teleological theory of ethics which sets as goal the greatest good (pleasure, benefit, etc.) of the one self alone. Egoism is derived from the Greek word "εγώ" (ego = myself). Egoism theory is distinguished in the following categories:

- *Psychological egoism* (based on the argument that humans are naturally motivated by self-interest).
- *Ethical egoism* (based on the argument that it is normative for the individuals to act in their own interest). The ethical egoist believes that whatever is for his/her own benefit is morally right.
- *Minimalist egoism* (better applied to social or economic processes where all agents are trying to get maximum profit with minimum loss). Clearly, this is neither a normative nor a descriptive approach.

Egoism theory is contrasted to *altruism* which is not restricted to the interests of one self alone, but includes in its goal the interest of others as well.

2.3.6 Value-Based Theory

The *value-based* theory uses some *value system* which consists of the ordering and prioritization of ethical and ideological values that an individual or community holds [11]. Value is *what* a person wants to do. It is not a deontological action but a *want-to-do* action. Two individuals or communities may have a set of common values but they may not have the same prioritization of them. Therefore, two groups of individuals with some of their values the same, may be in conflict with each other ideologically or physically. People with different value systems will not agree on the rightness or wrongness of certain actions (in general, or in specific situations).

Values are distinguished in [11]:

- *Ethical values* (which are used for specifying what is right or wrong, and moral or immoral). They define what is permitted or prohibited in the society that holds these values.
- *Ideological values* (which refer to more general or wider areas of religion, political, social and economic morals). A value system must be consistent, but in real-life this may not be true.

2.3.7 Case-Based Theory

This is a modern ethics theory that tries to overcome the apparently impossible divide between deontology and utilitarianism. It is also known as *casuistry* [12] and starts with immediate facts of a particular case. *Casuists* start with a particular case itself and then examine what are morally significant features (both theoretical and practical). Casuistry finds extensive application in juridical and ethical considerations of law ethics. For example, lying is always not permissible if we follow the deontological principle. However, in casuistry one might conclude that a person is wrong to lie in formal testimony under oath, but lying is the best action if the lie saves life.

2.4 Professional Ethics

Professional ethics provides guidance for interaction between professionals such that they can serve both each other and the whole society in the best way, without the fear of other professionals undercutting them with less ethical actions [13, 14]. Such codes are available in most professions, and are different from moral codes which are used to the education and religion of an entire larger society. Ethical codes are more specialized than moral codes, more internally consistent, and typically simple to be applied by an ordinary practitioner of the profession, without the need for extensive interpretation.

One of the earliest codes of professional ethics was the *Hippocrates Oath,* which provided rules for physician's ethical performance so as not to harm their patients. The oath and the whole code is written in the first person [15]. This medical profession code of ethics was revised by *Percival* [16] who defined acceptable conduct taking away the subjectivity of Hippocratic code.

Percival's code does not use the first person further discouraging personal interpretations of the code, and helps for a more consistent interpretation by different individuals, so that the standards are more universally applied [16]. This code was used as the basis for the formulation of professional ethics codes by many scientific and professional societies. Modern professional codes have the same attributes, specifying what primary duties and to whom a professional has, as unambiguously as possible.

In the following we present the following codes:

- The code of the National Society of Professional Engineers **(NSPE)** [17].
- The code of the Institute of Electrical and Electronic Engineers **(IEEE)** [18].
- The code of the American Society for Mechanical Engineers **(ASME)** [19].
- The code for robotics engineers developed by the Worcester Polytechnic Institute **(WPI)** [20].

2.4.1 NSPE Code of Ethics of Engineers

This code is stated as follows:

"Engineers in the fulfillment of their professional duties, shall:

1. Hold paramount the safety, health, and welfare of the public.
2. Perform services only in areas of their competence.

3. Issue public statements only in an objective and truthful manner.
4. Act for each employer or client as faithful agents or trustees.
5. Avoid deceptive acts.
6. Conduct themselves honorably, responsibly, ethically, and lawfully so as to enhance the honor, reputation, and usefulness of the profession".

This code is addressed to the entire engineering profession with no reference to particular engineering specialties. The detailed code which includes: (i) Rules of Practice, (ii) Professional Obligations, and (iii) a Statement by the NSPE Executive Committee, can be found in [17].

2.4.2 IEEE Code for Ethics

This code has ten attributes of ethical commitment and is stated as follows [18a]:

"We, the members of the IEEE, in recognition of the importance of our technologies in affecting the quality of life throughout the world, and in accepting a personal obligation to our profession, its members and the communities we serve, do hereby commit ourselves to the highest ethical and professional conduct and agree:

1. To accept responsibility in making decisions consistent with the safety, health and welfare of the public, and to disclose promptly factors that might endanger the public or the environment;
2. To avoid real or perceived conflicts of interest whenever possible, and to disclose them to affected parties when they do exist;
3. To be honest and realistic in stating claims or estimates based on available data;
4. To reject bribery in all its forms;
5. To improve the understanding of technology, its appropriate application, and potential consequences;
6. To maintain and improve our technical competence and to undertake technological tasks for other only if qualified by training or experience, or after full disclosure of pertinent limitations;
7. To seek, accept, and offer honest criticism of technical work, to acknowledge and correct errors, and to credit properly the contributions of others;
8. To treat fairly all persons regardless of such factors as race, religion, gender, disability, age, or national origin;
9. To avoid injuring others, their property, reputation, or employment by false or malicious action;
10. To assist colleagues and co-workers in their professional development and to support them in following this code of ethics".

Clearly, this code is again very general aiming to provide ethical rules for all electrical and electronic engineers. The IEEE code for conduct (of IEEE members and employees), approved and issued in June 2014, is given in [18b].

2.4.3 ASME Code of Ethics of Engineers

This code covers the entire profession of engineers and is formulated as follows [19]:

ASME requires ethical practice by each of its members and has adopted the following *Code of Ethics of Engineers*.

"*The Fundamental Principles*

Engineers uphold and advance the integrity, honor and dignity of the engineering profession by:

1. Using their knowledge and skill for the enhancement of human welfare;
2. Being honest and impartial, and serving with fidelity their clients (including their employers) and the public; and
3. Striving to increase the competence and prestige of the engineering profession".

"*The Fundamental Canons*

1. Engineers shall hold paramount the safety, health and welfare of the public in the performance of their professional duties.
2. Engineers shall perform services only in the areas of their competence; they shall build their professional reputation on the merit of their services and shall not compete unfairly with others.
3. Engineers shall continue their professional development throughout their careers and shall provide opportunities for the professional and ethical development of those engineers under their supervision.
4. Engineers shall act in professional matters for each employer or client as faithful agents or trustees, and shall avoid conflicts of interest or the appearance of conflicts of interest.
5. Engineers shall respect the proprietary information and intellectual property rights of others, including charitable organizations and professional societies in the engineering field.
6. Engineers shall associate only with reputable persons or organizations.
7. Engineers shall issue public statements only in an objective and truthful manner and shall avoid any conduct which brings discredit upon the profession.
8. Engineers shall consider environmental impact and sustainable development in the performance of their professional duties.
9. Engineers shall not seek ethical sanction against another engineer unless there is good reason to do so under relevant codes, policies and procedures governing that engineer's ethical conduct".

The detailed criteria for interpretation of the Canons are presented in [19].

2.4.4 *WPI Code of Ethics for Robotics Engineers*

This code is specialized to robotics engineers and is formulates as follows [20]:

"As an ethical robotics engineer, I understand that I have responsibility to keep in mind at all times the well being of the following communities:

Global—the good of people and the environment
National—the good of the people and government of my nation and its allies
Local—the good of the people and environment of affected communities
Robotics Engineers—the reputation of the profession and colleagues
Customers and End-Users—the expectations of the customers and end-users
Employers—the financial and reputation well-being of the company

To this end and to the best of my ability I will:

1. Act in such a manner that I would be willing to accept responsibility for the actions and uses of anything in which I have a part in creating.
2. Consider and respect people's physical well being and rights.
3. Not knowingly misinform, and if misinformation is spread do my best to correct it.
4. Respect and follow local, national, and international laws whenever applicable.
5. Recognize and disclose any conflicts of interest.
6. Accept and offer constructive criticism.
7. Help and assist colleagues in their professional development and in following this code".

As stated in [20], "this code was written to address the current state of robotics engineering and cannot be expected to account for all possible future developments in such a rapidly developing field. It will be necessary to review and revise this code as situations not anticipated by this code need to be addressed". Detailed discussions on robotic ethics and the WPI code of ethics for robotics engineers can be found in [21, 22], and a useful discussion on ethics and modular robotics is provided in [23].

2.5 Concluding Remarks

In this chapter we have presented the fundamental concepts and theories of ethics. The study of ethics in an analytical sense was initiated by the Greek philosophers Socrates, Plato and Aristotle who have developed what is called "*ethical naturalism*". Modern Western philosophers have developed other theories falling within the framework of analytic philosophy, which were described in the chapter. Actually, it is commonly recognized that there is an essential difference between ancient ethics and modern morality. For example, there appears to be a vital difference between *virtue theory* and the modern moralities of *deontological ethics*

(Kantianism) and *consequential ethics* (utilitarianism). But actually we can see that both ethical approaches have more in common than their stereotypes may suggest. Understanding the strengths and weaknesses of *virtue ethics* and *modern ethics* theories can help to overcome present-day ethical problems and develop fruitful ethical reasoning and decision-making approaches. The dominating current approach that individuals or groups follow in their relations is the *contract ethics* which is an implementation of minimalist theory. In *contract ethics* goodness is defined by mutual agreement for mutual advantage. This approach is followed because the players have more to gain than not.

References

1. Rachels J (2001) The elements of moral philosophy. McGraw-Hill, New York
2. Shafer-Landau R (2011) The fundamentals of ethics. Oxford University Press, Oxford
3. Singer P (2001) A companion to ethics. Blackwell Publishers, Malden
4. Singer P (1986) Applied ethics. Oxford University Press, Oxford
5. Hursthouse R (2002) On virtue ethics. Oxford University Press, Oxford
6. Ess C (2013) Aristotle's virtue ethics. Philosophy and Religion Department, Drury University. www.andrew.cmu.edu/course/80-241/guided_inquiries/more_vcr_html/2200.html, www.iep.utm.edu/virtue, www.drury.edu/ess/reason/Aristotle.html
7. Darwall S (ed) (2002) Deontology. Blackwell, Oxford
8. Gilty T (1951) St. Thomas aquinas philosophical texts. Oxford University Press, Oxford
9. Stuart Mill, J (1879) Utilitarianism, 7th edn. Longmans Green & co., London
10. Rawls J (1971) Theory of justice. The Belknap Press of Harvard University Press, Boston
11. Dewey J (1972) Theory of valuation. Chicago University Press, Chicago
12. Jonsen A, Toulmin S (1990) The abuse of casuistry: a history of moral reasoning. The University of California Press, Los Angeles
13. Beabout GR, Wennenmann DJ (1993) Applied professional ethics. University of Press of America, Milburn
14. Rowan JR, Sinaich Jr S (2002) Ethics for the professions. Cengage Learning, Boston
15. North T (2002) The Hippocratic Oath. National Library of Medicine. www.nlm.nih.gov/hmd/greek/greek_oath.html
16. Percival T (1803) Medical ethics. S. Russel, Manchester
17. NSPE: National Society of Professional Engineers: Code of Ethics for Engineers. www.nspe.org/Ethics/CodeofEthics/index.html
18. IEEE Code of Ethics and Code of Conduct. (a) www.ieee.org/about/corporate/governance/p7-8.html; (b) www.ieee.org/ieee_code_of_conduct.pdf
19. ASME Code of Ethics of Engineers. https://community.asme.org/colorado_section/w/wiki/8080.code-of-ethics.aspx
20. WPI Code of Ethics for Robotics Engineers (2010) http://ethics.iit.edu/ecodes/node/4391
21. Ingram B, Jones D, Lewis A, Richards M (2010) A code of ethics for robotic engineers. In: Proceedings of 5th ACM/IEEE international conference on human-robot interaction (HRI'2010), Osaka, Japan, 2–5 Mar 2010
22. www.wpi.edu/Pubs/E-project/Available/E-project-030410-172744/unrestricted/A_Code_of_Ethics_for_Robotics_Engineers.pdf
23. Smyth T, Paper discussing ethics and modular robotics. University of Pittsburgh www.pitt.edu/~tjs79/paper3.docx

Chapter 3
Artificial Intelligence

> *Instead of worrying about whether a particular machine can be intelligent, it is far more important to make a piece of software that is intelligent.*
>
> Oliver Selfridge

> *No computer has ever been designed that is ever aware of what it is doing; but most of the time, we aren't either.*
>
> Marvin Miasky

3.1 Introduction

The field of *artificial intelligence* (**AI**) is now over five decades old. Its birth took place at the so-called Dartmouth Conference (1956). Over these decades AI researchers have brought the field to a very high-level of advancement. The motivation for the extensive research in AI was the human dream to develop machines that are able to "*think*" in a human like manner and possess higher intellectual abilities and professional skills, including the capability of correcting themselves from their own mistakes. The computer science community is still now split into two schools of thought. The first school believes in AI and argues that soon AI will approach the ideal of human intelligence, why not surpass it. The second school argues against AI and believes that it is impossible to create computers (machines) that act intelligently.

For example, *Hubert Dreyfus* argued in 1979 that computer simulation workers assume incorrectly that explicit rules can govern intellectual processes. One of his arguments was that computer programs are inherently "*goal seeking*" and thus require the designer to know beforehand exactly what behavior is desired, as in chess match (as opposed to a work of art). In contrast humans are "*value seeking*", i.e., we don't always begin with an end goal in mind but seek to bring implicit values to fruition, the fly, through engagement in a creative or analytical process [1].

Alan Turing, John McCarthy, Helbert Simon and Allen Newell belong to the School arguing for AI. *Turing* states: "In the future there would be a machine that

S.G. Tzafestas, *Roboethics*, Intelligent Systems, Control and Automation: Science and Engineering 79, DOI 10.1007/978-3-319-21714-7_3

would duplicate human intelligence". *McCarthy* says: "Every aspect of learning or any other feature of intelligence can in principle be so precisely described that a machine can be made to simulate it", *Simon* argues: "Machines will be capable of doing any work Man can do", and *Newell* and *Simon* claim: "A physical symbol system has the necessary and sufficient means for general intelligent action". Moreover, many authors propose that the probability of *superintelligent software agent* being created within the coming few decades is high enough to warrant consideration [2, 3]. According to *Bostrom* [4], "Superintelligence is any intellect that vastly outperforms the best human brains in practically every field, including scientific creativity, general wisdom, and social skills". *Posner* [5] warns that as a superintelligence agent would have such enormous consequences, even if the probability of development is considered very low, the expected impact justifies careful consideration well before the capabilities to produce such an agent are realized. The ethical issues related with the development and use of intelligent computers (machines) are discussed in computer or machine or artificial intelligence ethics. These issues concern the cognitive and mental operations of the computer. Roboethics deals with both the intelligence and mechanical capabilities of the robots.

This chapter provides background material on AI which will help the reader in studying the ethics of AI and robotics discussed in Chaps. 5–9. Especially, the chapter:

- Discusses the differences between human and artificial intelligence.
- Discusses the Turing test that a machine must pass in order to be characterized as intelligent.
- Provides a tour to applied artificial intelligence.

3.2 Intelligence and Artificial Intelligence

The *artificial intelligence* (**AI**) field is the branch of computer science concerned with intelligent machines, or rather with embedding intelligence to computers. According to *McCarthy*, who coined the term in 1956, "artificial intelligence is the science and engineering of making intelligent machines" [6]. *Elaine Rich* (1983) defines artificial intelligence as "the branch of computer science that studies how we can make computers capable to do things that presently humans do better". This definition avoids the difficulty of defining the philosophical notions of "*artificial*" and "*intelligence*" [7]. The AI concept has brought about countless discussions, arguments, disagreements, misunderstandings and wrong hopes.

The *Central Intelligence Agency* defines the term intelligence as: "a collection of data and a computation of knowledge". This definition supports the proponents of AI. A statement that supports the opponents of AI is that of *Roger Penrose*: "true intelligence cannot be presented without *consciousness,* and hence intelligence can never be produced by any algorithm that is executed on a computer [8]".

The cultural definition of AI is: "the science and engineering of how to get machines to do things they do in the movies" such as the Stars of the Wars, Knight Rider, Star Trek: The Next Generation, The Matrix, Terminator, etc. These depictions of AI are not based on real science, but on the imagination of their screenwriters and science fiction authors.

Etymologically, the word *intelligence* comes from the Latin word "*intellegere*" (to understand, to perceive, to recognize and to realize). The first part of the word is derived from the prefix "*inter*" (which means "*between*"), and the second part comes from the word *legere* (meaning to select, to choose, and to gather). The combination of these two words can be interpreted as the capability to establish abstract links between details that have not any obvious relationship. Many researchers define "intelligence" as "*problem solving*" ability, but this is not correct. Knowing all facts and rules, and having access to every piece of information is not sufficient to provide *intelligence*. The essential part of intelligence (as the Latin word suggests) is the ability to look beyond the simple facts and givens, to capture and understand their connections and dependencies, and so to be able to produce new abstract ideas. Human beings do not only utilize their intelligence to solve problems. This is just one area where intelligence is applied. Very often human mind is in fact focused on some problem and works to solve it analytically or on the basis of past experience or both. But many other times one lets memories and observations drift through his/her mind like slow-moving clouds. This is also a form of thinking, although it is not a problem solving process and is not consciously directed at any goal. It is actually a *dreaming process* [10]. Thus understanding intelligence and thought must also include dreaming. In general, intelligence is used to coordinate and master the life, it is reflected in our behavior, and it motivates us towards achieving our goals, which are mainly derived by intelligence as well.

3.3 The Turing Test

The *Turing test* was coined by Turing [9] as a means to evaluate whether a machine is acting or thinking humanly. Applying this test one actually can give an answer to the question "*Can machines think?*" In this test a human interrogator presents a series of questions to a human being and to a machine. If the interrogator cannot tell which responses come from the machine, then the machine is considered to be acting humanly. Today no computer has yet passed this test in its full generality, and many scientists believe that none ever will [8]. Actually, a computer is given a set of instructions, but cannot understand more than a formal meaning of symbols. All who are not familiar with the details of the program are led to believe that the computer is working intelligently. Only the creators of the program know exactly how the algorithm is going to respond. IBM has developed a supercomputer called *Watson* who competed against humans on the game show *Jeopardy* [10], and a super computer named *Deep Blue* who was able to "*outthink*" the World Chess Champion *Garry Kasparov* with two wins and three draws (11 May 1997) [11].

But this is far from the required proof that a computer can truly think like a human. A deep examination of how the above performance of the supercomputer in the Jeopardy and chess games it was achieved, reveals that computers so far are better than humans in two specific functions:

- The ability to store and retrieve enormous amount of data in order to solve a specific problem.
- The superiority in calculation.

Once a computer has searched and retrieved possible solutions, it computes the relevancy via algorithms and available data sets to come up to a suitable answer.

For example, in the chess game Deep Blue, the computer was able to search future moves to a depth of "between six and eight moves to a maximum of twenty, or more in some cases" [12]. But, though the computer has this disproportional search advantage, it does not have the ability to create "*strategy*" which is a crucial ability to go beyond mere chess moves. The lack of strategy is replaced by using a brute force approach of predicting moves. A second reason why computers cannot be truly intelligent is their inability to "*learn independently*". Humans solve problems based on previous experiences and other cognitive functions, but accumulate those memories via *independent learning*.

Concerning the human and computer intelligence *Weizenbaum* states: "there is an aspect of the human mind, the '*unconscious*', that cannot be explained by the information-processing primitives, the elementary information processes, which we associate with formal thinking, calculation, and systematic rationality" [13]. About the same issue *Attila Narin* argues: "there is no way that a computer can ever exceed the state of simply being a data processor dealing with zeros and ones, doing exactly what it is told to do. Electronic machinery will never reach the point where a machine has a life or could be called intelligent. Only tools that superficially mimic intelligence have been created [14].

The human thinking process involves a spectrum of "*focus levels*" from the upper (analytical) end with the highest focus level, via the "mind-drifting relaxation levels" with intermediate focus levels, to the "dreaming" level with the lowest "focus level". *Gelernter* [15] parallelizes the human focus spectrum with the light spectrum where the upper end is the highest frequency ultraviolet color and the lower end is the infra-red lowest frequency color. The intermediate violet, blue, green yellow, orange, red colors have decreasing frequencies like the decreasing thinking focus levels of the human thought. As one grows tired during the day, he/she becomes less capable of analytical high-focus thought and more likely sits back and drifts. Actually during the day we oscillate many times, from higher to lower focus and then back to higher focus, until we drop off at last in sleep. Our current focus level is an aggregation of several physiological factors. The above discussion suggests that the "*thinking focus*" must be incorporated in any view of the mind. The primary goal of theoretical AI is exactly the exploration and understanding of the cognitive spectrum, which leads to an understanding of *analogy* and *creativity*. Only a few steps towards this goal have so far been completed. A contributed book on the theoretical foundations of *artificial general intelligence* (**AGI**) is [16], where novel solutions to a repertory of theoretical AI problems are presented in a coherent way, linking AGI research with research in allied fields.

3.4 A Tour to Applied AI

Despite the theoretical and philosophical concerns about whether artificial intelligence can reach or exceed the capabilities of human intelligence, artificial intelligence has become an important element of computer industry helping to solve extremely difficult problems of society. Applied AI is concerned with the implementation of intelligent systems, and so is a type of engineering or at least an applied science. Actually, applied AI must be regarded as a part of engineering, implementing the intelligent systems for the following:

- Natural language understanding and processing
- Speech understanding and processing
- Mechanical/computer vision
- Autonomous/intelligent robots
- Domain expertise acquisition

Of course the initial systems for theorem proving and game-playing are included in Applied AI as well as the subfields of *knowledge engineering* (**KE**) and *expert systems* (**ES**). The term "knowledge engineering" was coined in 1983 by Feigenbaum and McCorduck [17] as the engineering field that integrates knowledge into computer systems in order to solve complex problems normally requiring high-level and professional expertise. Each ES deals with a specific problem domain requiring high-level and professional expertise. Thus, the knowledge in ESs is acquired from human experts. In *knowledge based systems* (**KBS**) the knowledge is derived by mathematical simulation models or drawn from real experimental work. Actually an ES simulates the human reasoning by using specific rules or objects representing the human expertise. The fundamental steps in the development of AI were done in the decades 1950–1980 and are shown in Table 3.1.

Table 3.1 Evolution of AI in the period 1950–1980

Decade	Area	Researchers	Developed system
1950	Neural networks	Rosenblatt (Wiener, McCulloh)	Perceptron
1960	Heuristic search	Newell and Simon (Turing, Shannon)	GPS (General problem solving)
1970	Knowledge representation	Shortliffe (Minsky, McCarthy)	MYCIN
1980	Machine learning	Lenat (Samuel, Holland)	EURISCO

Due to the broad spectrum of human application areas and problems which are dealt with by AI, the approaches of solution required are numerous and quite different from each other. However, there are some basic methods that play major roles in all cases. These are the following [18, 19]:

- Knowledge acquisition and maintenance
- Knowledge representation
- Solution search
- Reasoning
- Machine learning

The most important and popular methods of knowledge representation are the following:

- Predicate logic (calculus)
- Propositional calculus
- Production rules
- Semantic networks
- Frames and scripts
- Objects
- Model-based representation
- Ontologies

The last method which is based on the "*ontology*" concept is relatively newer than the other methods, and we will discuss it in short here. The term "ontology" was borrowed from philosophy, where ontology is a branch of metaphysics that deals with the study of *being* or *existence* and λόγος (logos = study). *Aristotle* has described ontology as "*the science of being*". *Plato* has considered that ontology is related to "*ideas*" and "*forms*". The three concepts that play a dominant role in metaphysics are: "*substance*", "*form*", and "*matter*".

In knowledge engineering the term "ontology" is used as a "representation" of knowledge in knowledge bases. Actually, any ontology offers a shared vocabulary that can be used to model and represent the type of objects or concepts of a domain, i.e., it offers a formal explicit specification of a shared conceptualization [20]. In practice most ontologies represent individuals, classes (concepts), attributes and relations. An ES designed using the "ontology method" is PROTÉGÉ II.

The principal ways *for solution search* in the state space of AI problems are [19]:

- Depth-first-search
- Breadth-first search
- Best-first search

All of them belong to the so-called *generate-and-test* approach in which the solution are generated and subsequently tested in order to check their match with the situation at hand.

Reasoning with the stored knowledge is the process of drawing conclusions from the facts in the knowledge base, or, actually, inferring conclusion from premises.

The three classical ways of knowledge representation directly understandable by the computer are:

- Logic expressions
- Production rules
- Slot-and-filler structures

A form of reasoning which is known as *automated reasoning* is very successfully employed in expert systems. This is based on *logic programming* and implemented in the PROLOG language.

Machine learning is an AI process difficult to define uniquely, since it ranges from the addition of any single fact or a new piece of new knowledge to complex control strategy, or a proper rearrangement of system structure, and so on. A useful class of machine learning is the *automated learning* which is the process (capability) of an intelligent system to enhance its performance through learning, i.e., by using its previous experience. Five basic automated learning aspects are:

- Concept learning
- Inductive learning (learning by examples)
- Learning by discovery
- Connectionist learning
- Learning by analogy

A complete expert system involves the following four basic components [21, 22]:

- The knowledge base
- The inference engine
- The knowledge acquisition unit
- The explanation and interface unit

A knowledge-based system may not have all the above components. The initial expert systems were built using the higher-level languages available at that time. The two higher-level languages most commonly used in the past for AI programming are LISP and PROLOG. At the next level, above the higher-level programming, are programming environments designed to help the programmer/designer to accomplish his/her task. Other, more convenient programs for developing expert systems are the so-called "*expert system building tools*", or "*expert systems shells*" or just "*tools*". The available tools are distinguished in the following five types [23]:

- Inductive tools
- Simple rule-based tools
- Structured rule-based tools
- Hybrid tools
- Domain specific tools
- Object-oriented tools

A subfield of applied AI is the area of human-computer interaction which is performed using appropriate *human-computer interfaces* (**HCI**s) [24–26]. The HCI field has been shown to have a significant influence on factors such as learning time, speed of performance, error rates, and human user's satisfaction. While good interface design can lead to significant improvements, poor designs may hold the user back. Therefore, work on HCIs is of crucial importance in AI applications. HCIs are distinguished in two broad categories:

- Conventional HCIs
- Intelligent HCIs

The conventional HCIs include the keys and keyboards and the pointing devices (touch screens, light pens, graphic tablets, trackballs, and mice) [24]. Intelligent HCIs are particularly needed in automation and robotics. First, there are increasing levels of data processing, information fusion, and intelligent control between the human user/operator and the real system (plant, process, robot, enterprise), and source of sensory data. Second, advanced AI and KB techniques are needed to be employed and embedded in the loop to achieve high-levels of automation via monitoring, event detection, situation identification, and action selection functions. The HCI should be intelligent in the sense that has access to a variety of knowledge sources such as knowledge of the user tasks, knowledge of the tools, knowledge of the domain, knowledge of the interaction modalities, and knowledge of how to interact [26].

3.5 Concluding Remarks

In this chapter we have considered a set of fundamental concepts of AI which allow the reader to appreciate the discussion of its ethical and societal issues rising in robotics. The ethical issues of AI arise from the belief that intelligent (or super-intelligent) computers, which may drive autonomous processes, decisions and robot actions, is feasible to design and built.

In any case even the present day AI capabilities, that simulate many of the human intelligence capabilities, suggest a serious consideration of its impact on society. A few representative societal areas where AI has the potential to affect human life are medicine, assistive technology, house keeping technology, safety, driverless transport, weather forecast, business, economic processes, etc. In the future the human-machine interdependence will be strengthened such that to achieve shared goals that neither could achieve alone. In most cases the human will be setting the goals, formulate the hypotheses, determine the criteria, and carry out the evaluations. The computers will mainly do the routine-type of work required for the decisions. Current analysis indicates that the human-computer symbiotic partnership will perform intellectual operations much more efficiently than human alone can perform [27].

References

1. Dreyfus HL (1979) What computers can do: the limits of artificial intelligence. Harper Colophon Books, New York
2. Hall JS (2001) Beyond AI: Creating the Conscience of the Machine. Prometheus Books, Amherst
3. Moravec H (1999) Robot: mere machine to transcendent mind. Oxford University Press, New York
4. Bostrom N (2003) Ethical issues in advanced artificial intelligence. In: Smit I, Lasker GE (eds) Cognitive, emotive and ethical aspects of decision making in humans and artificial intelligence, vol 2. International Institute for Advanced Studies in Systems Research/Cybernetics, Windsor, ON, pp 12–17
5. Posner R (2004) Catastrophe risk and response. Oxford University Press, New York
6. McCarthy J, Hayes PJ (1969) Some philosophical problems from the stand point of artificial intelligence. In: Meltzer B, Michie B (eds) Machine intelligence, 4th edn. Edimburgh University Press, Edimburgh, pp 463–502
7. Rich E (1984) Artificial intelligence. Mc Graw-Hill, New York
8. Noyes JL (1992) Artificial intelligence with common lisp. D.C. Heath, Lexington
9. Turing A (1950) Computing machinery and intelligence. MIND 49:433–460
10. "What is Watson", IBM Innovation, IBM Inc. www.ibm.com/innovation/us/Watson/what-is-watson/index.html
11. Hsu F-H (2002) Behind deep blue: building the computer that defeated the world chess champion. Princeton University Press, Princeton
12. Campbell M (1998) An enjoyable game. In: Stork DG (ed) HAL's legacy: 2001 computer as dream and reality. MIT Press, Cambridge
13. Weizenbaum J (1976) Computer power and human reason. W.H. Freeman, California
14. Narim A (1993) The myths of artificial intelligence. www.narin.com/attila/ai.html
15. Gelernter D, What happened to theoretical AI? www.forbes.com/2009/06/18/computing-cognitive-consciousness-opinions-contributors-artificial-intelligence-09-gelernter.html
16. Wang P, Goertzel B (eds) (2012) Theoretical foundations of artificial general intelligence. Atlantis Thinking Machines, Paris
17. Feigenbaum EA, McCorduck P (1983) The fifth generation. Addison-Wesley, Reading, MA
18. Barr A, Feigenbaum EA (1971) Handbook of artificial intelligence. Pittman, London
19. Popovic D, Bhatkar VP (1994) Methods and tools for applied artificial intelligence. Marcel Dekker, New York/Basel
20. Gruber T (1995) Towards principles for the design of ontologies used for knowledge sharing. Int J Hum Comput Stud 43(5–6):907–928
21. Forsyth R (1984) Expert systems. Chapman and Hall, Boca Raton, Fl
22. Bowerman R, Glover P (1988) Putting expert systems into practice. Van Nostrand Reinhold, New York
23. Harman P, Maus R, Morrissey W (1988) Expert systems: tools and applications. John Wiley and Sons, New York/Chichester
24. Lewis J, Potosnak KM, Mayar RL (1997) Keys and keyboards. In: Helander MG, Landawer TK, Prabhu P (eds) Handbook of human-computer interaction. North-Holland, Amsterdam
25. Foley JD, van Dam A (1982) Fundamentals of interactive computer graphics. Addison-Wesley, Reading, MA
26. Tzafestas SG, Tzafestas ES (2001) Human-machine interaction in intelligent robotic systems: a unifying consideration with implementation examples. J Intell Rob Syst 32(2):119–141
27. Licklider JCR (1960) Man-computer symbiosis. IRE Trans Hum Factors Electron HFE 1 (1):4–11

Chapter 4
The World of Robots

At bottom, robotics is about us. It is the discipline of emulating our lives, of wondering how we work.

Rod Grupen

From where I stand, it is easy to see the science lurking in robotics.
It lies in the welding of intelligence to energy. That is, it lies in intelligent perception and intelligent control of motion.

Allen Newel

4.1 Introduction

*Robotic*s lies at the crossroad of many scientific fields such us mechanical engineering, electrical-electronic engineering, control engineering, computer science and engineering, sensor engineering, decision-making, knowledge engineering, etc. It has been established as a key scientific and technological field of modern human society, having already offered considerable services to it.

The development of robots into "*intelligent machines*" will offer further opportunities to be exploited, new challenges to be faced, and new fears to be examined and evaluated. Future intelligent robots will be fully autonomous multiarm mobile machines, capable to communicate via human-robot natural-like languages, and to receive, translate, and execute general instructions. To this end, new developments in metafunctional sensing, cognition, perception, decision making, machine learning, on-line knowledge acquisition, reasoning under uncertainty, and adaptive and knowledge-based control will have to be embodied.

The objective of this chapter is to provide a tour to the world of robots including the entire gamma of robots from industrial to service, medical, and military robots. This will be used in the ethical considerations to be discussed in the book. In particular the chapter:

S.G. Tzafestas, *Roboethics*, Intelligent Systems, Control and Automation: Science and Engineering 79, DOI 10.1007/978-3-319-21714-7_4

- Explains what is a robot and discusses the types of robots by kinematic structure and locomotion.
- Provides a short discussion of intelligent robots that employ AI techniques and tools.
- Presents the robot applications in industry, medicine, society, space, and military.

Purposefully, the material of this chapter has a pure descriptive (non technical) nature which is deemed to be sufficient for the purposes of the chapter.

4.2 Definition and Types of Robots

4.2.1 Definition of Robots

The word *robot* (*robota*) was coined by the Chzech dramatist *Karel Capel* in his play named "Rossum's Universal Robots" [1]. In the drama the "robot" is a humanoid which is an intellectual worker with feelings, creativity and loyalty. The robot is defined as follows: "Robots are not people. They are mechanical creatures more perfect than humans, they have extraordinary intellectual features, but they have not soul. The creation of engineers are technically more refined than the creations of nature".

Actually, the first robot in the worldwide history (around 2500 BC) is the Greek mythodological creature called *Talos (Τάλως)*, supernature science-fiction man with bronze body and a single vein from the neck up to the ankle in which the so-called "*ichor*" (the blood of the immortals) was flowing. Around 270 BC the Greek engineer *Ktesibios* (Κτησίβιος) has designed the well known "water clock" and around 100 AD *Heron of Alexandria* designed and constructed several feedback mechanisms such as the odometer, the steam boiler, the automatic wine distributor and the automatic opening of Temples. Around 1200 AD the Arab author *Al Jazari* has written "Automata" which is one of the most important texts in the study of the history of technology and engineering. Around 1490 *Leonardo Da Vinci* constructed a device that looks as an armored Knight which is considered to be the first humanoid (android) robot in the Western civilization. In 1940 the science fiction writer *Isaak Asimov* used for the first time the terms "robot" and "robotics" and stated his three laws of robotics, known as *Asimov's laws* which we be discussed later in the book.

The actual start of modern robotics is the year 1954 when *Devol Jr* patented his multi joined robotic arm. The first industrial robot named *Unimate* was put in operation by the company *Unimation* (**Uni**versal Auto**mation**) in 1961.

Actually, there is not a global or unique scientific definition of a robot. The Robotics Institute of America (RIA) defines an industrial robot as: "a reprogrammable multi-functional manipulator designed to move materials, parts, tools, or specialized devices through variable programmed motions for the performance of a variety of tasks which also acquire information from the environment and move intelligently in response". This early definition does not capture mobile robots. The European Standard EN775/1992 defines the robot as: "Manipulating industrial robot is an automatically controlled reprogrammable multipurpose, manipulative machine with several degrees of freedom, which may be either fixed in place or mobile for use in industrial automation applications". *Ronald Arkin* gave the following definition: "An intelligent robot is a machine able to extract information from its environment and use knowledge about its work to move safely in a meaningful and purposive manner".

In general, an intelligent robot is referred to as a machine that performs an intelligent connection between *perception* and *action*, and an autonomous robot is a robot that can work without human intervention, and with the aid of embodied artificial intelligence can perform and live within its environment.

4.2.2 Types of Robots

The evolution of robots after the appearance of the Unimate robot in 1961 showed an enormous expansion both in the structured and the applications Landmarks in this expansion are: the *Rancho Arm* (1963), the *Stanford Arm* (1963), the mobile robot *Shakey* (SRI Technology, 1970), the *Stanford Cart* (1979), The Japanese humanoid robots *WABOT-1* (1980) and *WABOT-2* (1984), the *Waseda-Hitachi Log-11* (1985), the *Aqaurobot* (1989), the multi-legged robot *Genghis* (1989), the exploring robot *Dante* (1993) and *Dante II* (1994), the NASA *Sojourner* robotic rover (1997), the HONDA humanoid *ASIMO* (2000), the FDA *cyberknife* for treating human tumors (2001), *SONY AIBOERS-7,* third generation *robotic pet* (2003), the SHADOW *dexterous hand* (2008), the Toyota running *humanoid robot* FLAME (2010), and many others.

In terms of geometrical structure and locomotion the robots are classified as [2, 3]:

- Fixed robotic manipulators
- Wheeled mobile robots
- Biped robots (humanoids)
- Multilegged robots
- Flying robots
- Undersea robots
- Other

Fixed robotic manipulators This class involves the following types of robots: Cartesian, cylindrical, spherical (polar), articulated, SCARA, parallel, and gantry robots (Fig. 4.1). Cartesian robots have three linear (translational) axes of motion, cylindrical robots have two linear and one rotational axis, spherical robots have one linear and two rotational axes, articulated robots have three rotating axes, SCARA (Selective Compliance arms for Robotic Assembly) are a combination of cylindrical

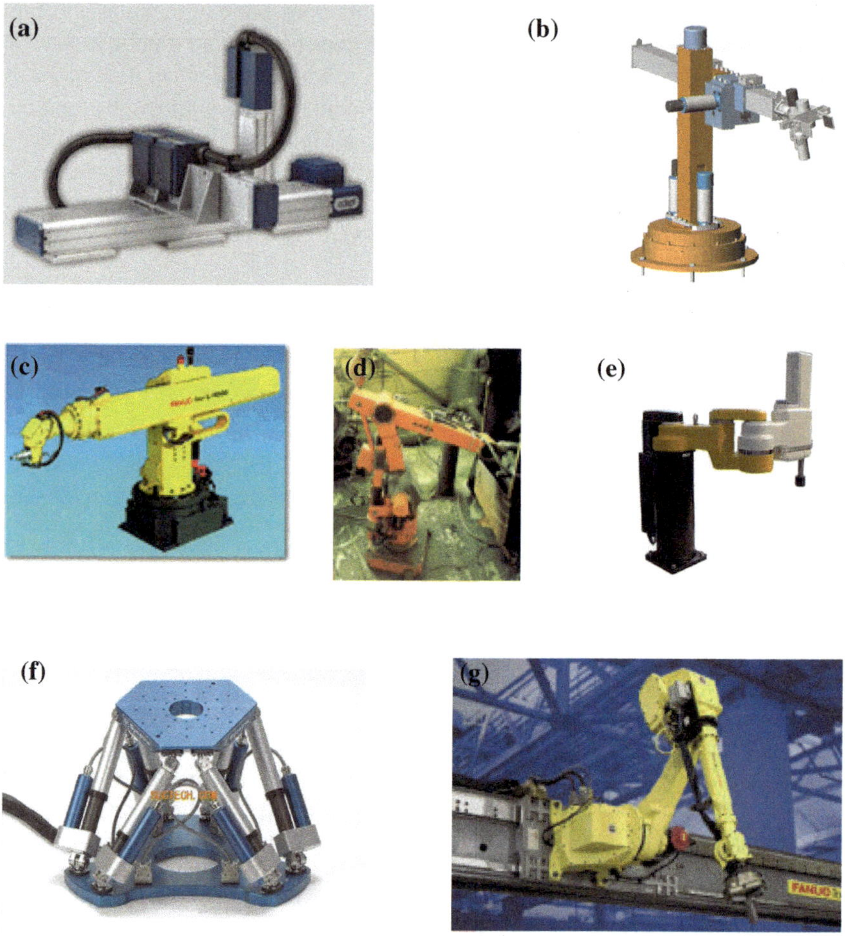

Fig. 4.1 Representative industrial fixed robots, **a** cartesian, **b** cylindrical, **c** spherical, **d** articulated (anthropomorphic), **e** SCARA (selective compliance arm for robotic assembly), **f** parallel, **g** gantry robot. *Source* (**a**) http://www.adept.com/images/products/adept-python-3axis.jpg, (**b**) http://robot.etf.rs/wp-content/uploads/2010/12/virtual-lab-cylindrical-configuration-robot.gif, (**c**) http://www.robotmatrix.org/images/PolarRobot.gif, (**d**) http://www02.abb.com/global/gad/gad02007.nsf/Images/F953FAF81F00334DC1257385002D98BF/$File/IRB_6_100px.jpg, (**e**) http://www.factronics.com.sg/images/p_robot_scara01.jpg, (**f**) http://www.suctech.com/my_pictures/of_product_land/robot/parallel_robot_platform/6DOF5.jpg, (**g**) http://www.milacron.com/images/products/auxequip/robot_gantry-B220.jpg

and articulated robots, parallel robots use Steward platforms, and gantry robots consist of a robot arm mounted on an overhead track, creating a horizontal plane that the robot can travel, thus extending the work envelope.

Wheeled mobile robots These robots move using two types of wheels: (i) conventional wheels, and (ii) special wheels [4]. *Conventional wheels* are distinguished in powered fixed wheels, castor wheels, and powered steering wheels. Powered fixed wheels are driven, by motors mounted on fixed positions of the vehicle. Castor wheels are not powered and can rotate freely about an axis perpendicular to their axis of rotation. Powered steering wheels have a driving motor for their rotation and can be steered about an axis perpendicular to their axis of rotation. *Special* wheels involve three main types: universal wheel, mecanum wheel, and ball wheel. The universal wheel contains small rollers around its outer diameter which are mounted perpendicular to the wheel's rotational axis. This wheel can roll in the direction parallel to the wheel axis in addition to the wheel rotation. The mecanum wheel is similar to the universal wheel except that the rollers are mounted at an angle different than 90°, (typically ±45°). The ball wheel (or spherical) wheel can rotate to any direction providing an omnidirectional motion to the vehicle.

According to their drive type, wheeled mobile robots are distinguished in: (i) differential drive robots, (ii) tricycle robots, (iii) omnidirectional robots (with universal, mecanum, synchrodrive, and ball wheels), (iv) Ackerman (car like) steering, and (v) skid steering, Fig. 4.2 shows the photos of typical wheeled robots of the above types.

Biped robots These robots have two legs like the human and move in three modes, namely: (i) standing on the two legs, (ii) walking, and (iii) running.

Fig. 4.2 Representative real mobile robots **a** pioneer 3 differential drive, **b** tricycle, **c** Ackerman (car like) drive, **d** omnidirectional (universal drive), **e** omnidirectional (mecanum drive), **f** omnidirectional (synchro drive), **g** skid steering (tracked robot). *Source* (**a**) http://www.conscious-robots.com/images/stories/robots/pioneer_dx.jpg, (**b**) http://www.tinyhousetalk.com/wp-content/uploads/trivia-electric-powered-tricycle-car-concept-vehicle.jpg, (**c**) http://sqrt-1.dk/robot/img/robot1.jpg, (**d**) http://files.deviceguru.com/rovio-3.jpg, (**e**) http://robotics.ee.uwa.edu.au/eyebot/doc/robots/omni2-diag.jpg, (**f**) http://groups.csail.mit.edu/drl/courses/cs54-2001s/images/b14r.jpg, (**g**) http://thumbs1.ebaystatic.com/d/l225/m/mQVvAhe8gyFGWfnyolgtRjQ.jpg

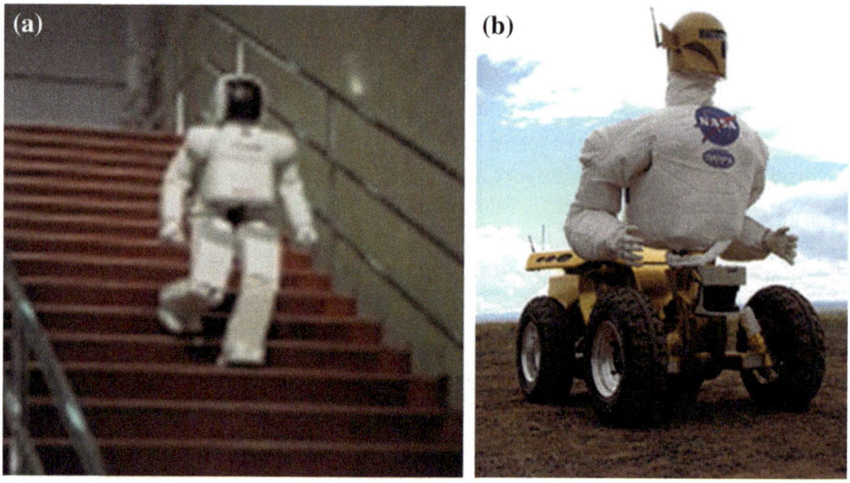

Fig. 4.3 Examples of humanoids, **a** HONDA ASIMO humanoid. **b** NASA Robonaut. *Source* (**a**) http://images.gizmag.com/gallery_tn/1765_01.jpg, (**b**) http://robonaut.jsc.nasa.gov/R1/images/centaur-small.jpg

Humanoid robots (humanoids) are biped robots with an overall looking based on that of the human body, i.e., they have head, torso, legs, arms and hands [5] (Fig. 4.3a). Some humanoids may model, only part of the body, e.g., waist-up like NASA's Robonaut, while others have also a "face" with "eyes" and mouth.

Multilegged robots The original research on multilegged robots was focused on robot locomotion design for smooth and easy rough terrain, for passing simple obstacles, body maneuvering, motion on soft ground, and so on. These requirements can be realized via periodic gaits and binary (yes/no) contact information from the ground. Newer studies are concerned with multi-legged robots that can move over an impassable road or an extremely difficult terrain such as mountain areas, ditches, trenches, and damage areas from earthquakes, etc. A basic issue in this research is to guarantee the robot stability under very difficult ground conditions [6]. Two examples of multilegged robots are shown in Fig. 4.4.

Flying robots These robots include all robots that can move in the air either under the control of a human pilot or autonomously. The former involve all types of aircrafts, helicopters, including aerostats. The autonomously guided aircrafts and vehicles, called *unmanned aerial vehicles* **(UAVs)**, are typically used for military purposes [7]. Other flying robots usually used for entertainment purposes may have a bird-like or insect-like locomotion. Figure 4.5 shows six examples of flying robots.

Undersea robots These robots find important applications replacing human in undersea operations [8]. Working underwater is both difficult and dangerous for humans. Many robots have the capability of both swimming in the sea and walking on the seabed and the beach.

Fig. 4.4 Two typical examples of real multi-legged robots, **a** DARPA LS3 quadruped robot. **b** NASA/JPL six-legged spider robot. *Source* (**a**) http://images.gizmag.com/gallery_lrg/lc3-robot-26.jpg, **b**) http://robhogg.com/wp-content/uploads/2008/02/poses_02.jpg

Other robots have a fish like form, used also for interesting undersea operations. Still, others have a lobster looking. Figure 4.6 shows three examples of undersea robots.

Other robot types Besides the above land, sea and air robot types, roboticists have developed over the years various robots that mimic biology or combine wheels, legs and wings. Just to take an idea of what a robot might be we give in Fig. 4.7 an insect-like robot, a spinal robot, and a stair climbing robot.

4.3 Intelligent Robots: A Quick Look

Intelligent robots, i.e., the robots that have embedded artificial intelligence and intelligent semi-autonomous or autonomous control, constitute the class of robots that rise the most crucial ethical issues as discussed in the next chapters. Therefore, a quick look at them will be useful for the study of roboethics. Intelligent robots are specific types of machines that obey Saridis' principle of "*increasing intelligence with decreasing precision*" [9]. They have decision-making capabilities and use multisensory information about their internal state and work space environment to generate and execute plans for carrying out complex tasks. They can also monitor the execution of their plans, learn from past experience, improve their behavior, and communicate with a human operator in natural or almost natural language. The key feature of an intelligent robot (which is not possessed by a non-intelligent robot) is that it can perform a repertory of different tasks under conditions that may not be known a priori. All types of robots discussed in Sect. 4.2.2 can become (and in many ways have become) intelligent with embedded AI and intelligent control of various levels of intelligence. The principal components of any intelligent robotic system are: effectors (arms, hands, wheels, winds and legs with their actuators,

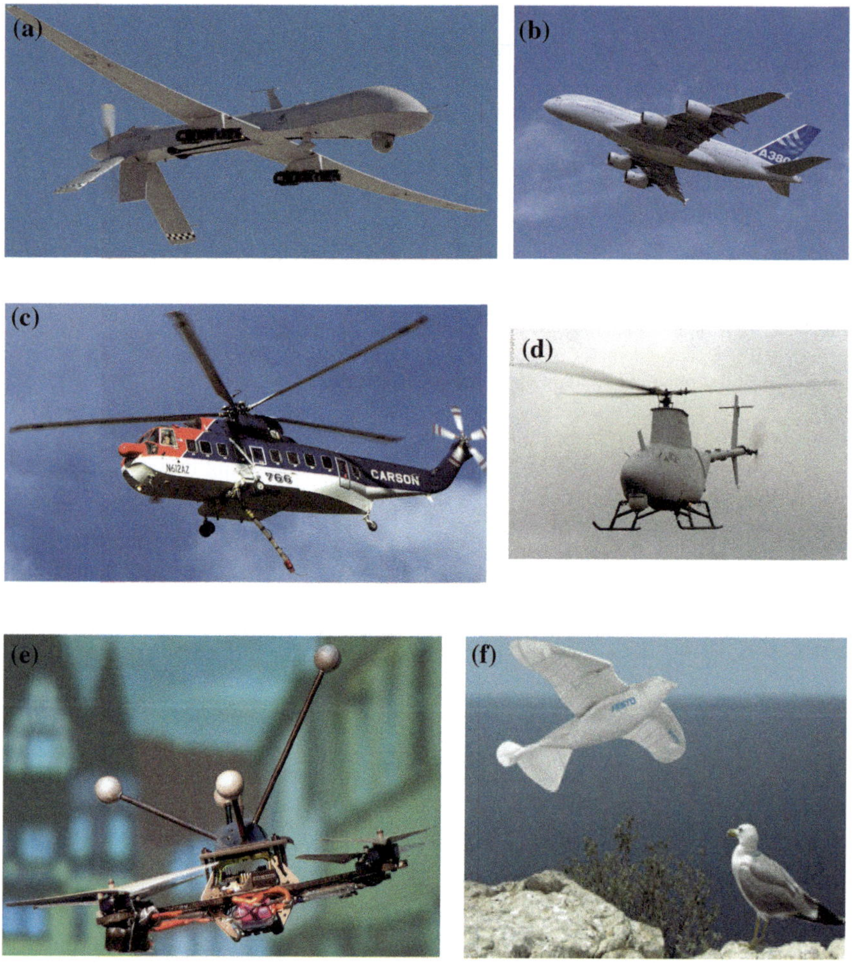

Fig. 4.5 Flying robots, **a** predator AUV, **b** commercial aircraft, **c** transfer Helicopter, **d** unmanned helicopter, **e** ayeron scout (flying camera), **f** robot bird. *Source* (**a**) http://www.goiam.org/uploadedImages/mq1_predator.jpg, (**b**) http://2.bp.blogspot.com/-tGTpv0F2KPc/Tdgnxj-36AI/AAAAAAAACis/suKRwhC0IUI/s1600/Airbus+Airplanes+aircraft.jpg, (**c**) http://blog.oregonlive.com/news_impact/2009/03/18373141_H12269589.JPG, (**d**) http://www.strangecosmos.com/images/content/175835.jpg, (**e**) http://tuebingen.mpg.de/uploads/RTEmagicP_Quadcopter_2_03.jpg, (**f**) http://someinterestingfacts.net/wp-content/uploads/2013/02/Flying-Robot-Bird-300x186.jpg

motors and propulsion mechanisms), sensors (acoustic, range, velocity, force or touch, vision, etc.), computers (local controllers, supervisors, coordinators, execution controllers, and auxiliary equipment (end-arm tools, pallets, fixtures, platforms, etc.) [10–12].

In general, intelligent robots can perform several variants of the following operations:

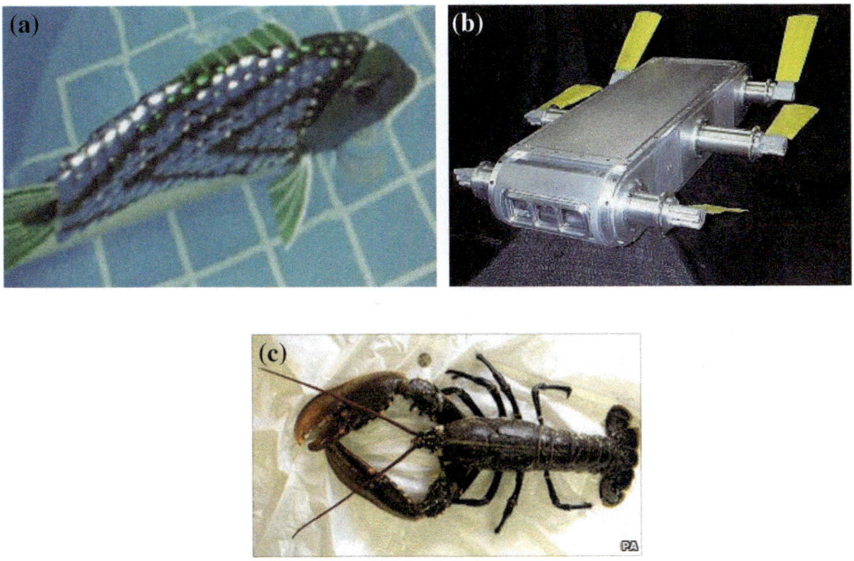

Fig. 4.6 **a** A robotic fish used for monitoring water pollution, **b** the AQUA robot that can take pictures of coral reefs and other aquatic organisms, **c** a lobster-like robot (autonomous). *Source* (**a**) http://designtopnews.com/wp-content/uploads/2009/03/robotic-fish-for-monitoring-water-pollution-5.jpg, (**b**) http://www.rutgersprep.org/kendall/7thgrade/cycleA_2008_09/zi/OverviewAQUA.JPG, (**c**) http://news.bbcimg.co.uk/media/images/51845000/jpg/_51845213_lobster,pa.jpg

- Cognition
- Perception
- Planning
- Sensing
- Control
- Action

The above operations are combined in several ways into the specific system architecture which is adopted in each case and the actual control actions and tasks required. A general architecture for structuring the above operations of intelligent robots (IRs) is shown in Fig. 4.8.

The computer of the *Intelligent robotic systems* **(IRS)** communicates (and interacts) with the surrounding world and performs the *cognition, perception, planning* and *control* functions. The computer also sends information to the robot under control and receives information provided by the sensors. The cognition is needed for the organization of the repertory of information obtained from the sensors that may have quite different physical types. Usually a database and an inference engine is employed, not only for the interpretation of the cognition results, but also for putting them in the proper order which is needed for the determination of the strategies of the future robot operation and the planning and control actions. The purpose of the planner/controller is to generate the proper control sequences that are needed for the successful control of the robot.

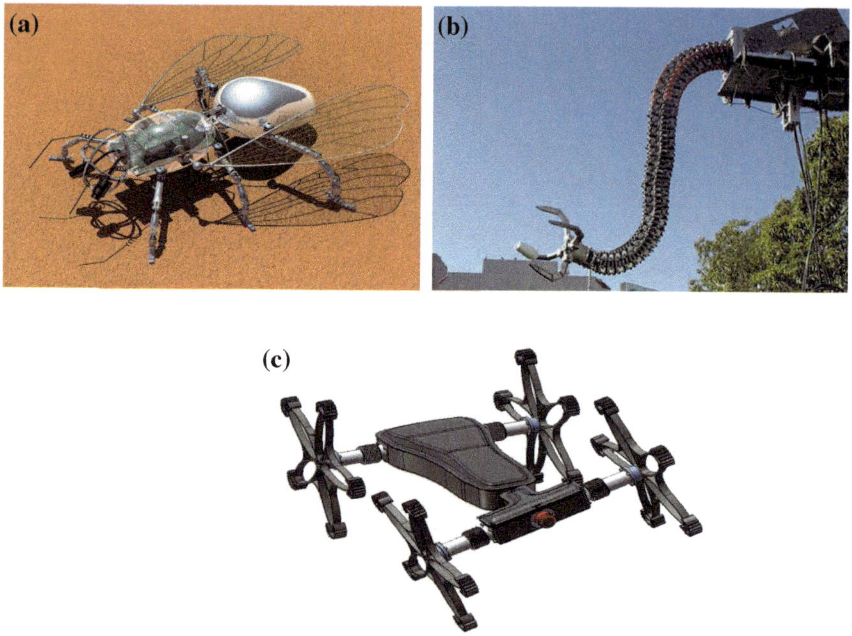

Fig. 4.7 Three robot examples, **a** insect-like robot, **b** spinal robot, **c** stair climbing robot. *Source* (**a**) http://assets.inhabitat.com/wp-content/blogs.dir/1/files/2011/11/Cyborg-Insect-1-537x392.jpg#Insects%20%20537x392, (**b**) http://techtripper.com/wp-content/uploads/2012/08/Spine-Robot-1.jpg, (**c**) http://www.emeraldinsight.com/content_images/fig/0490360403006.png

Fig. 4.8 General architecture of intelligent robotic systems

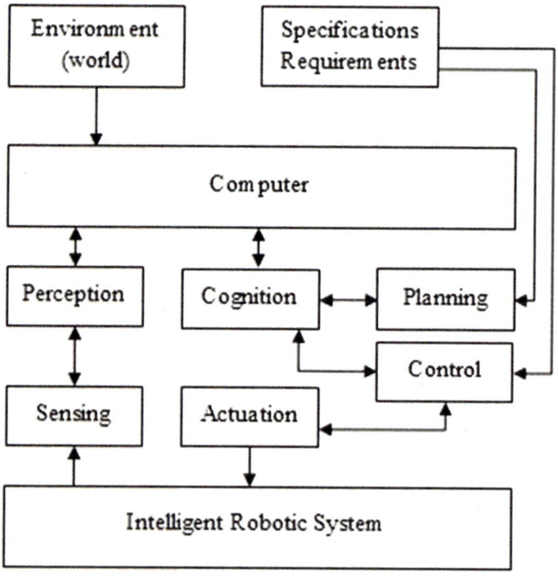

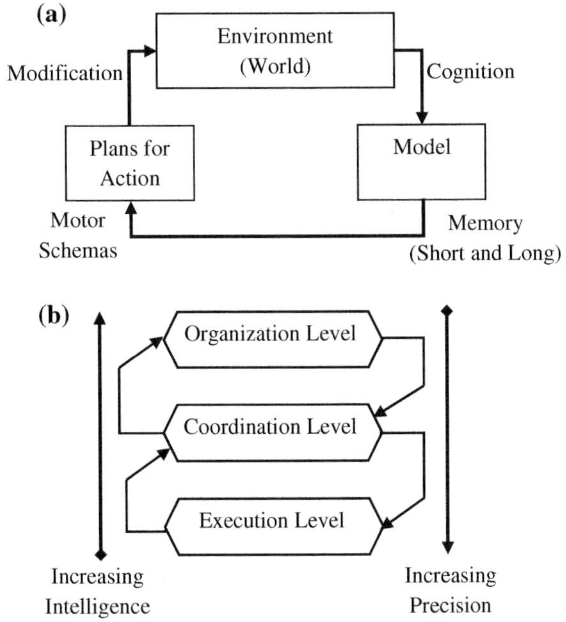

Fig. 4.9 **a** The perception-action cycle. **b** The three hierarchical levels of intelligent robotic systems

The intelligent robots perform autonomously the following four tasks:

- Obstacle avoidance
- Goal recognition
- Path/motion planning
- Localization and mapping (with the aid of landmarks).

It can be easily seen that most of the cognition tasks can be decomposed in two distinct phases: (i) *recognition*, and (ii) *tracking*. The recognition is primarily based on predictions/estimations that are produced by the internal models of the landmarks. The joint operation and information exchange between the recognition, the perception and the action (motion) is represented successfully by the so-called "perception-action" of intelligent robot control, which is shown in Fig. 4.9a.

The robotic systems are actually classified in the following categories:

- **Non autonomous robotic systems** These systems need a control processor for the execution of the *offline* and *online* computations.
- **Semi-autonomous robotic systems** These systems react independently to variations of the environment computing new path sections in real time.
- **Autonomous robotic systems** These systems require supervision from some internal coordinator and use the work plans that are generated by themselves during their operation.

The intelligent robotic systems belong to the class of hierarchical intelligent systems which have three main levels according to Saridis' principle (Fig. 4.9b). These levels are the following:

- Organization level
- Coordination level
- Execution level

The *organization level* receives and analyzes the higher-level commands and performs the higher-level operations (learning, decision making, etc.). It also gets and interprets feedback information from the lower levels.

The *coordination level* consists of several coordinators, each one being realized by a software or hardware component.

The *execution level* involves the actuators, the hardware controllers, and the sensing devices (sonar, visual, etc.), and executes the action programs issued by the coordination level.

The control problem of an intelligent robotic system is split in two subproblems:

- The logic or functional control subproblem that refers to the coordination of the events under the restrictions in the sequence of events.
- The geometric or dynamic control subproblem which refers to the determination of the geometric and dynamic motion parameters such that all geometric, dynamic constraints and specifications are satisfied.

Other architectures proposed for intelligent robotic control (besides Saridis hierarchical architecture) include [4, 11, 12]:

- Multiresolutional architecture (*Meystel*)
- Reference model architecture (*Albus*)
- Subsumption behavior based architecture (*Brooks*)
- Motor schemas behavior based architecture (*Arkin*).

4.4 Robot Applications

So far we have presented the classification of robots by geometric structure and locomotion, and how robots get AI features, i.e., we have discussed how robots can do things. Here, we will present the classification of robots by their applications, i.e., by looking at the things they are doing or can do. A convenient robot classification by applications is the following:

- Industrial robots
- Medical robots
- Domestic and household robots
- Assistive robots
- Rescue robots
- Space robots
- Military robots
- Entertainments robots.

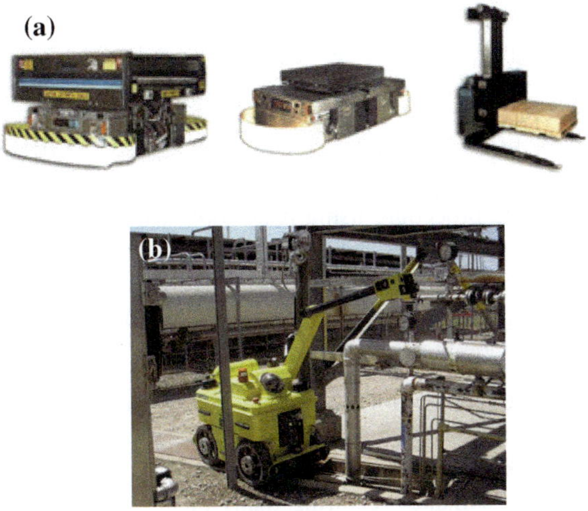

Fig. 4.10 **a** Three Corecon AGVs (conveyor, platform, palletizing box lamp), **b** an industrial mobile manipulator. *Source* (**a**) http://www.coreconagvs.com/images/products/thumbs/thumbR320.jpg, http://www.coreconagvs.com/images/products/thumbs/thumbP320p.jpg, http://www.coreconagvs.com/images/products/thumbs/thumbP325.jpg, (**b**) http://www.rec.ri.cmu.edu/about/news/sensabot_manipulation.jpg

4.4.1 Industrial Robots

Industrial robots or *factory robots* represent the larger class of robots, but a big "boom" was shown during the last two decades in the medical and social applications of intelligent robots. The dominant areas of robotics in the factory are the following [13]:

- Machine loading and unloading
- Material handling and casting
- Welding and assembly
- Machining and inspection
- Drilling, forging, etc.

Mobile robots are used in factory for material handling and product transfer from one place to another for inspection, quality control, etc. Figure 4.10 shows examples of *autonomous guided vehicles* (**AGVs**) for material transfer.

4.4.2 Medical Robots

Robots are used in the medical field for hospital material transport, security and surveillance, floor cleaning, inspection in the nuclear field, explosives handling, pharmacy automation systems, integrated surgical systems, and entertainment. Surgical robotic systems represent a particular class of telerobots that allow surgeons to perform dexterous surgical operations much more accurately and successfully than classical surgery [13, 14]. The field of robotic assisted surgery was originally developed for military use. In classical surgery, the surgeon formulates a general diagnosis and surgical plan, makes an incision to get access to the target anatomy, performs the procedure using hand tools with visual or tactile feedback, and closes the opening. Modern anesthesia, sterility methods, and antibiotics have made classical surgery extremely successful. However, the human has several limitations that brought this classical approach to a point of diminishing returns. These limitations include the following:

- It is still hard to couple medical imaging information (X-rays, CT, MRI, ultra-sound, etc.) to the surgeon's natural hand-eye coordination (limited planning and feedback).
- Natural hand tremor makes repairing many anatomical structures (e.g., retina, small nerves, small vascular structures) extremely tedious or impossible (limited precision).
- The dissection needed to gain access to the target is often far more traumatic than the actual repair. Minimally invasive methods (e.g., endoscopic surgery) provide faster healing and shorter hospital stays, but severely limit the surgeon's dexterity, visual feedback, and manipulation precision (limited access and dexterity).

The above difficulties can be overcome by automated surgery via computer-integrated robotic surgical systems. These systems exploit a variety of modern automation technologies such as robots, smart sensors and human-machine interfaces, to connect the "*virtual reality*" of computer models of the patient to the "*actual reality*" of the operating room.

One of the most popular surgical robots is *DaVinci robot* a 7-degreed-of-freedom robot which, with the aid of a 3D visual display, provides the surgeon with an enhanced capability allowing him/her to perform *minimally invasive surgery* **(MIS)** more precisely [14, 15]. Minimal invasive surgery (also known as laparoscopic or endoscopic surgery) is performed by making 1–3 cm incisions, and using pencil-sized tools that are passed through natural body cavities and incisions. In traditional surgical operations the surgeons start with a single, long incision (about 6–12 inches) and visualize the site of interest by opening wide the body cavities. On the contrary laparoscopic surgery makes use of endoscopy (the insertion of cameras into the body cavity) to help the surgeon visualize the inside of the body without the need of long incisions. It is remarked that smaller incisions imply less blood-loss, pain or probability of infection, and reduces the recovery time from surgery. Medical/surgical robotic systems are distinguished in:

- *CASP/CASE* which integrate computer-assisted surgical planning (CASP) and computer-assisted surgical execution (CASE) operations with robots.

- *Surgical augmentation systems* which extend human sensory-motor abilities to overcome many of the limitations of classical surgery.
- *Surgical assistant systems* that work in a cooperative manner with a surgeon to automate many of the tasks performed by surgical assistants.

Certainly, robotic surgery is useful only in some cases. For example, patients undergoing an appendectomy are not gaining much by robotic surgery, whereas procedures done on the prostate have shown significant outcomes with the use of surgical robot. Figure 4.11 shows a snapshot of the robotic surgery using DaVincy robot, and Fig. 4.12 shows a hospital mobile service robot.

4.4.3 Domestic and Household Robots

Domestic and household robots are mobile robots and mobile manipulators designed for household tasks such as floor cleaning, pool cleaning, coffee making, serving, etc. Robots capable for helping elderly people and persons with special

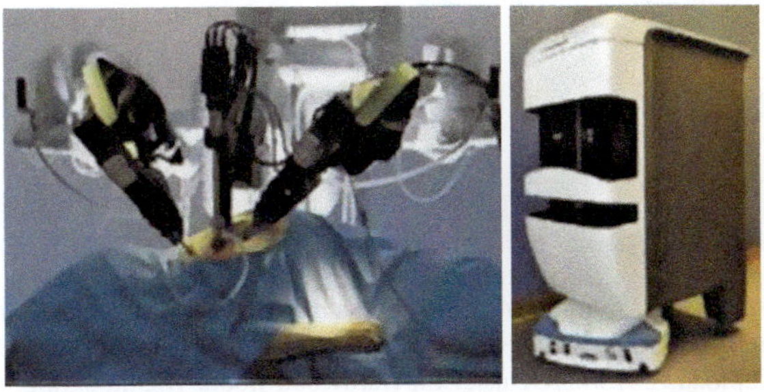

Fig. 4.11 The Da Vinci robot at work. *Source* https://encrypted-tbn1.gstatic.com/images?q=tbn:ANd9GcRnqQZo_kRHSDfdd3ta8n5lagxasG3o4pxO7nY3OvrYXHYeepUtOg

Fig. 4.12 A hospital mobile service robot.. *Source* http://www.hotsr.com/contents/uploads/pictures/2013/05/Hospital-robot.jpg

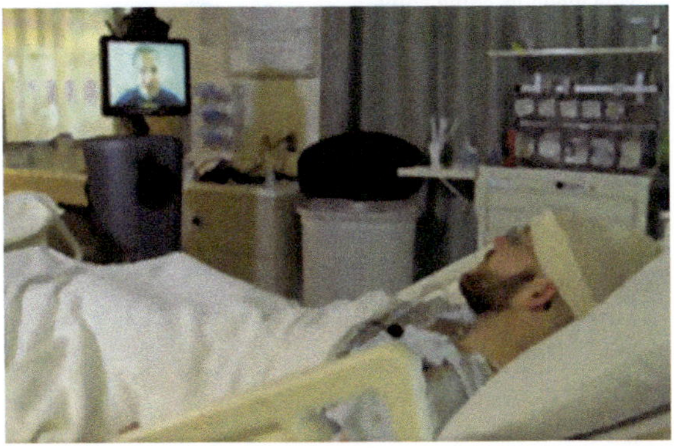

needs (PwSN) may also be included in this class, although they can be regarded to belong to the more specialized class of assistive robots. Today home robots include also humanoid robots suitable for helping in the house [16]. Examples of domestic and home robots are the following.

O-Duster robot This robot can clean tile, linoleum, hard wood, and other hard, floored surfaces. The soft edges of the flexible base allow the robot to reach easily all odd corners in the room (Fig. 4.13).

Swimming pool cleaner robot A robot that goes up to the bottom of the pool and after finishing the cleaning job returns on the surface (Fig. 4.14).

Care-O-bot 3 This is a robot that has a highly flexible arm with a three finger hand which is capable of picking up home items (a bottle, a cup, etc.). It can, for example, carefully pick-up a bottle of orange, juice and put it next to the glasses on the tray in front of it (Fig. 4.15). To be able to do this it is equipped with many sensors (stereo vision color cameras, laser scanners and a 3-D range camera).

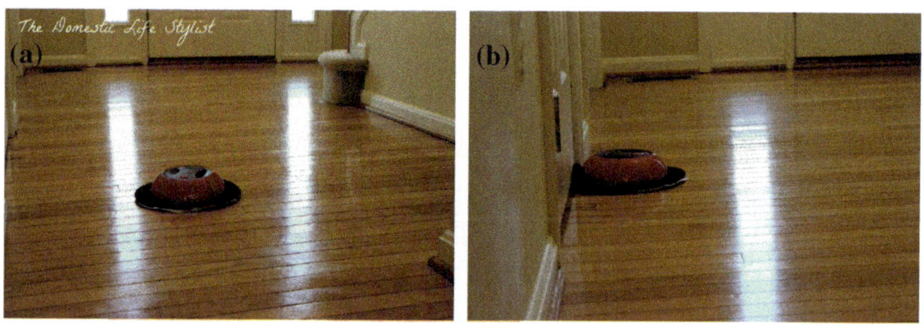

Fig. 4.13 O-duster cleaner robot at work cleaning a hardwood floor. **a** The robot is working in the middle of a room. **b** The robot reaches a wall of the room. *Source* (**a**) http://thedomesticlifestylist.com/wp-content/uploads/2013/03/ODuster-on-Hardwood-floors-1024x682.jpg, (**b**) http://fortikur.com/wp-content/uploads/2013/10/Cleaning-Wood-Floor-with-ODuster-Robot.jpg

Fig. 4.14 Swimming pool cleaner robot. *Source* http://image.made-in-china.com/2f0j00HKjTNcotEfkA/Swimming-Pool-Cleaner-Robot.jpg

Fig. 4.15 The Care-O-Bot omnidirectional home robot (height 1.4 m). *Source* http://www.flickr.com/photos/lirec/5839209614/in/photostream/

Dust cart and Dust clean The Dust cart mobile humanoid robot (Fig. 4.16a) is a garbage collector, and the Dust clean mobile robot can be used for automatic cleaning narrow streets (Fig. 4.16b).

4.4.4 Assistive Robots

Assistive robots belong to *assistive technology* (**AT**) which is in our times is a major field of research, given the ageing of population and diminishing number of available care givers. Assistive robotics includes all robotic systems that are developed for **PwSN** and attempt to enable disabled people to reach and maintain their best physical and/or social functional level, improving their quality of life and work productivity [17–19]. The people with special needs are classified as:

- PwSN with loss of lower limb control (paraplegic patients, spinal cord injury, tumor, degenerative disease)
- PwSN with loss of upper limb control (and associated locomotor disorders)

(a) (b)

Fig. 4.16 **a** The humanoid garbage collector "Dust cart". **b** The mobile robot street cleaner "Dust clean". *Source* (**a**, *left*) http://www.greenecoservices.com/wp-content/uploads/2009/07/dustcart_robot.gif, (**b**, *right*) http://www.robotechsrl.com/images/dc_image011.jpg

- PwSN with loss of spatio-temporal orientation (mental, neuropsychological impairments, brain injuries, stroke, ageing, etc.)

Today many smart autonomous robots are available for PwSN, including:

(i) *Smart-intelligent wheelchairs* that can eliminate the user's task to drive the wheelchair and can detect and avoid obstacles and other risks.
(ii) *Wheelchair mounted manipulators* (WMMs) which offer the best solution for people with motor disabilities increasing the user's mobility and the ability to handle objects, and perform everyday functions. Today WMMs can be operated in all alternative ways (manual, semi-automatic, automatic) through the use of proper interfaces.
(iii) *Mobile autonomous manipulators* (MAMs), i.e., robotic arms mounted on mobile platforms, that can follow the user's (PwSN's) wheelchair in the environment, can perform tasks in open environments, and can be shared between several users.

Figure 4.17 shows two robotic wheelchair systems designed for helping PwSPs.

4.4.5 Rescue Robots

Natural and manmade disaster offer unique challengers for effective cooperation of robots and humans. The locations of disasters are usually too dangerous for human intervention or cannot be reached. In many cases there are additional difficulties

Fig. 4.17 Two autonomous wheelchairs. **a** A wheelchair with mounted a service robotic manipulator. **b** A carrier robotic wheelchair that can traverse all terrains including stair climbing. *Source* (**a**) http://www.iat.uni-bremen.de/fastmedia/98/thumbnails/REHACARE-05.jpg.1725.jpg, (**b**) http://www.robotsnob.com/pictures/carrierchair.jpg

such as extreme temperatures, radioactive levels, strong wind forces, etc. that do not allow a fast action of human rescuers Lessons learned from past disaster experience have motivated extended research and development in many countries for the construction of suitable robotic rescuers. Due to the strong earthquake activity Japan is one of the countries where powerful and effective autonomous or semi-autonomous robotic systems for rescue were developed. Modern robot rescuers are light flexible and durable. Many of them have cameras with 360° rotation that provide high resolution images, and other sensors that can detect body temperature and colored clothing. Figure 4.18 shows two examples of rescue robots.

Fig. 4.18 Two examples of robots for rescue. *Source* (**a**) http://www.technovelgy.com/graphics/content07/rescue-robot.jpg, (**b**) http://www.terradaily.com/images/exponent-marcbot-us-and-r-urban-search-rescue-robot-bg.jpg

4.4.6 Space Robots

The applications of robots in the outer space are very important and were deeply studied over the years in a series of research programs at NASA and elsewhere (Germany, Canada, etc.). Robots that function efficiently in space present several unique challenges for engineers. Most types of lubricants that are used on earth cannot be used in space because of the ultra vacuum conditions held there. There is no gravity, a fact that permits to create several unique systems. The temperature conditions in the robot vary tremendously depending on whether the robot is in the sun light or shade. The subfield of robotics developed for the space an other remote applications on the ground or in the deep sea is called *telerobotics* [20, 21]. Telerobots combine the capabilities of standard robots (fixed or mobile) and teleoperators. Teleoperators are operated by direct manual control and needs an operator to work in real time for hours. Of course due to the human supervision they can perform non repetitive tasks (as, e.g. it is required in nuclear environments). A telerobot has more capabilities than either a standard robot or a teleoperator, because it can carry out many more tasks that can be accomplished by each one of them alone. Therefore the advantages of both are fruitfully exploited, and their limitations minimized.

NASA has put a considerable research effort and investment in three fundamental areas [22]:

- Remote operations on planetary and lunar surfaces.
- Robotics tending of scientific payloads.
- Satellite and space system servicing.

These areas required advance automation technology (to reduce crew interaction), hazardous material handling, robotic vision systems, collision avoidance algorithms, etc.

One of the first NASA spacecrafts for conducting scientific studies on the surface of another planet was *Viking* (called *Viking Lander*). Viking 1 was orbiting Mars (1997) with its robotic rover called *Sojourner* (Fig. 4.19).

A more recent *Mars Exploration Rover* (MER), called the *phoenix MER* was sent to Mars on August 4, 2007, to investigate the existence of water and life-supporting conditions on Mars. The Canadian space telerobot, named Canadarm, was designed such that to be able to help astronauts to "toss" satellites into space and then collect the faulty ones. Canadarm can be equipped with different dexterous end-effectors that allow the astronauts to perform high precision tasks. One of them is shown in Fig. 4.20.

Figure 4.21 shows an advanced space rover equipped with a large number of exploration sensors for atmospheric, surface and biological experiments.

Fig. 4.19 NASA path finder Viking. *Source* http://www.nasa.gov/images/content/483308main_aa_3-9_history_viking_lander_1024.jpg

Fig. 4.20 A dexterous end-effector for the space telerobot Canadarm. *Source* http://www.dailygalaxy.com/photos/uncategorized/2008/03/10/080307dextre_robot_hmed_4pwidec_2.jpg

4.4.7 *Military Robots*

The design, development and construction of robots for the war has got a substantial portion of the investment in robotics research and application. In general, military robots operate in geopolitically sensitive environments. Autonomous war robots include missiles, unmanned combat air vehicles, unmanned terrestrial vehicles, and autonomous underwater vehicles [23, 24]. Military robots, especially lethal ones, rise the most critical ethical implications for the human society.

Fig. 4.21 A space robotic rover. *Source* http://4.bp.blogspot.com/-t6odDvdzXkk/ToXGnOauDCI/AAAAAAAABHE/5PaNDE7gFco/s400/Space-Robots.jpg

Terrestrial military robots Fig. 4.22 shows two *unmanned ground vehicles* **(UGV)** funded by DARPA and designed by Carnegie Mellon's National Robotics Engineering Center (NREC). This type of UGV is called *crusher* and is suitable for reconnaissance and support-tasks and can carry huge payloads including armor.

Figure 4.23 shows a terrestrial "modular advanced armed robotic system" (MAARs) which is capable of acting on the front lines, and is equipped with day and night cameras, motion detectors, acoustic microphone, and a speaker system.

Marine Military Robots Fig. 4.24 shows an unmanned mine-detection boat (called *Unmanned Influence Sweep System*: **UISS**) equipped with a magnetic and acoustic device that it tows. UISS can replace helicopters that are used to do this kind of mine sweeps.

Fig. 4.22 Two types of Crusher UGV. *Source* (**a**) http://static.ddmcdn.com/gif/crusher-3.jpg, (**b**) http://static.ddmcdn.com/gif/crusher-4.jpg

Fig. 4.23 Modular advance armed system. *Source* http://asset2.cbsistatic.com/cnwk.1d/i/tim2/2013/05/31/img_maars.jpg

Fig. 4.24 Unmanned influence sweep system. *Source* http://defensetech.org/2013/05/29/navy-developing-unmanned-mine-detection-boat/uiss/

Figure 4.25 shows an *autonomous underwater vehicle* **(AUV)** that can detect and neutralize/destroy ship hull mines and *underwater improvised explosive devices* **(IED)**.

Aerial military robots Figs. 4.26, 4.27 and 4.28 show some representative *unmanned aerial vehicles* **(UAV)**.

The *X-47B* is a kind of stealth UAV that closely resembles a strike fighter. It can take off from and land on an aircraft carrier and support mid-air refueling. It has a range of 3380 km, it can fly up to 40,000 feet at high subsonic speed up to 2000 kg ordnance in two weapon.

Patriot is a long-range, all-altitude, all-weather air defense system to counter tactical ballistic missiles, cruise missiles and advanced aircrafts.

The *laser-guided smart bomb* (Unit-27) is equipped with a computer, optical sensor, and a pattern. The control system steers the bomb so that the reflected laser, beam is hitting near the center of the photodiode array. This keeps the bomb heading toward the target.

Fig. 4.25 HAUV-N underwater IED detector and neutralizer (Bluefin Robotics Corp.). *Source* http://www.militaryaerospace.com/content/dam/etc/medialib/new-lib/mae/print-articles/volume-22/issue-05/67368.res

Fig. 4.26 The X-47B stealth UAV. *Source* (**a**, *left*) http://www.popsci.com/sites/popsci.com/files/styles/article_image_large/public/images/2013/05/130517-N-YZ751-017.jpg?itok=iBZ_S6Sl, (**b**, *right*) http://www.dw.de/image/0,,16818951_401,00.jpg

4.4.8 Entertainment and Socialized Robots

Entertainment and socialized robots are high-level intelligent autonomous robots equipped with sensors, able to learn social rules and performance. They are able to interact with humans in a humanistic way with every-day persons (nonprofessionals). Most of the available socialized and entertainment robots have a humanoid appearance, flexibility, and adaptability. Currently, there are educational and research entertainment socialized robots under development in several research institutes, and commercially available robots that can write, dance, play musical instruments, soccer, and interact emotionally. Educational robotic kits are designed

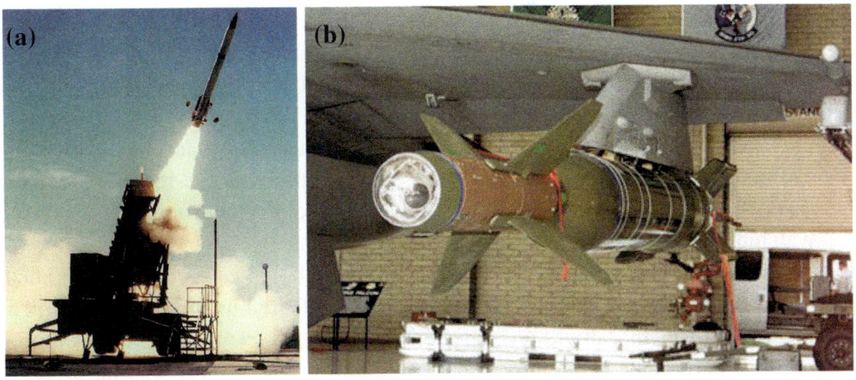

Fig. 4.27 **a** A Patriot missile being fired. **b** A laser guided smart bomb. *Source* (**a**) http://www.army-technology.com/projects/patriot/images/pat10.jpg, (**b**) http://static.ddmcdn.com/gif/smart-bomb-5.jpg

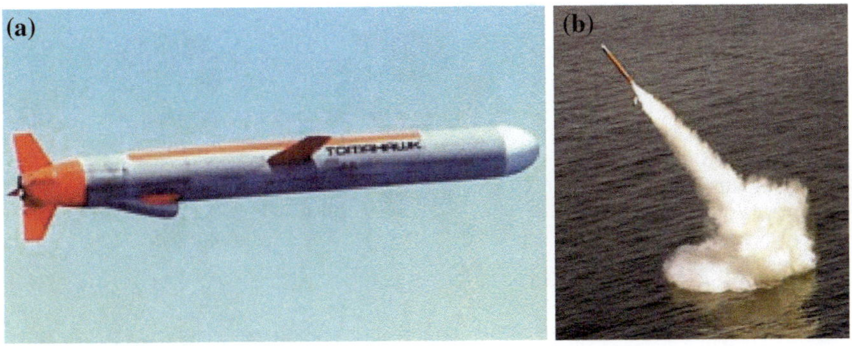

Fig. 4.28 The Tomahawk submarine-launched cruise missile. *Source* (**a**) http://static.ddmcdn.com/gif/cruise-missile-intro-250x150.jpg, (**b**) http://static.ddmcdn.com/gif/cruise-missile-launch-water.jpg

for fun and competitions where students put together modules in innovative ways to create robots that work. Humanoid robots and innovative shaped robots are increasingly taking a place in homes and offices. The modularity of robot kits makes them versatile and flexible [25, 26].

A partial list of basic social skills that an entertainment/socialized robot must have is the following [27]:

- Ability to contact with humans in a repeated and long-life setting.
- Ability to negotiate tasks and preferences and provide "*companionship*".
- Ability to become personalized, recognizing and adapting to its owner's preferences.

- Ability to adapt, to learn and expand its skills, e.g., by being taught new performances by its owner.
- Ability to play a role of companion in a more human-like way (probably similarly to pets).
- *Social skills*. These are essential for a robot to be acceptable as a companion. For example, it is good to have a robot that says "would you like me to being a cup of coffee?" But, it may not be desirable to ask this question while you are watching your favorite TV.
- When moving in the same area as a human, the robot always changes its route to avoid getting to close to the human, especially if the human's back is turned.
- The robot turns its camera properly to indicate by its gaze that it was looking in order to participate or anticipate what is going on in the surrounding area.

Figures 4.29 and 4.30 depict three representative entertainment robots.

Finally, Figs. 4.31 and 4.32 show three well known socialized robots (KASPAR, Kismet, and Sony bot AIBO).

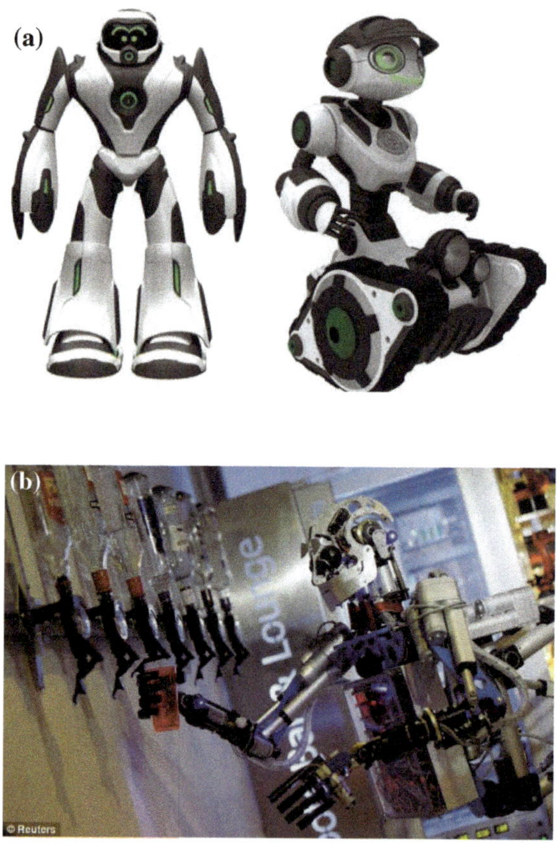

Fig. 4.29 Three entertainment robots, **a** wowweebots, **b** the Robo-Bartender CARL developed in the German company H & S Robots. *Source* (**a**) http://www.bgdna.com/images/stories/Robotics/wowweebots.jpg, (**b**) http://www.robotappstore.com/images/rasblog/Carl%20the%20robot%20bartender.jpg

Fig. 4.30 Humanoids in a soccer game. *Source* http://www.designboom.com/cms/images/-Z80/rob1.jpg

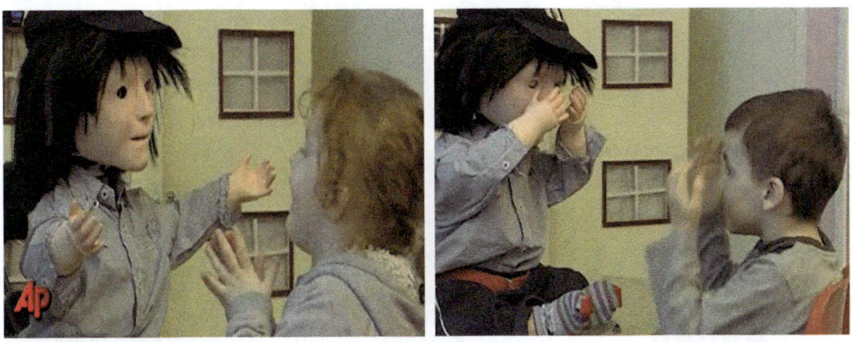

Fig. 4.31 KASPAR interacting with children. *Source* (**a**) https://lh3.googleusercontent.com/-Py8bkmZ7QR8/TXbNMwI9g3I/AAAAAAAAekQ/Kx-veky_zhg/s640/Friendly+Kid+Robot+Kaspar+vs.+Helps+Autistic+Children+Learning+Emotion+1.jpg, (**b**) https://lh4.googleusercontent.com/-uL7Q3NtzmL0/TXbNTytKHGI/AAAAAAAAekc/Ctmx5Tl6g_w/s640/Friendly+Kid+Robot+Kaspar+vs.+Helps+Autistic+Children+Learning+Emotion+4.jpg

KASPAR (Kinesis and Synchronization in Personal Assistant Robot) is a child sized humanoid robot developed at the University of Hertfordshire (UK).Using skin sensor technologies KASPAR gets cognitive mechanisms that use this tactile feedback for improving the human-robot interaction. It can play with children with autism who can explore social communication and interaction in a safe and enjoyable way [28].

Kismet was created at MIT (Media Lab.) and named from the Turkish word 'kismet' which means "destiny" or "fate". Kismet is an expressive robot with perceptual and motor capabilities. It can read both audio and visual social cues. The

Fig. 4.32 **a** The MIT socialized robot Kismet. **b** The socialized robotic dog AIBO. **c** Several expressions of Kismet. *Source* (**a**) http://web.mit.edu/museum/img/about/Kismet_312.jpg, (**b**) http://www.sony.net/SonyInfo/News/Press_Archive/199905/99-046/aibo.gif, (**c**) http://faculty.bus.olemiss.edu/breithel/final%20backup%20of%20bus620%20summer%202000%20from%20mba%20server/frankie_gulledge/artificial_intelligence_and_robotics/expressions-lips2.jpg

motor system provides vocalizations, facial expressions, and adjustment of the gaze direction of the eyes and the orientation of the head. It can also steer the visual and auditory sensors to the source of the stimulus, and display communicative cues. A full presentation of Kismet, together with generic features of sociable robots, is included in [29, 30].

AIBO (Artificial Intelligence roBOt) can be used as companion and adjunct to therapy for children with autism and elderly with dementia [31].

4.5 Concluding Remarks

This chapter has presented an outline of the basic concepts of robotics which will help in the discussion of the ethical issues of robotics to be given in the next chapters. The class of robots with strong labor implications is the class of industrial robots which do not actually need much intelligence and autonomy. The one with the strongest ethical concerns is the class of autonomous land-air-sea robotic weapons. The robots that also rise challenging ethical questions are the surgical robots and the therapeutic socialized robots. Proponents of autonomous robotic weapons try to explain that these weapons behave in the battlefield more ethically than human-controlled weapons. Opponents of their use argue that autonomous lethal weapons are completely not acceptable and must be prohibited. Surgical robots have shown to enhance the quality of surgery in many cases, but they complicate the ethical and legal liability in case of malfunctioning. Assistive robots are subject to all medical ethical rules with emphasis on the selection of the most proper device that surely helps the user to do things he/she finds hard to do. Finally, socialized robots especially those that are used for children socialization and elderly companionship are subject to the emotional attachment of the user to the robot, the lessening of the human care, the user awareness, and the user's privacy ethical concerns, as explained in Chap. 8.

References

1. Freedman J (2011) Robots through history: robotics. Rosen Central, New York
2. Angelo A (2007) A reference guide to new technology. Greenwood Press, Boston, MA
3. McKerrow PK (1999) Introduction to robotics. Wesley, Reading, MA
4. Tzafestas SG (2013) Introduction to mobile robot control. Elsevier, New York
5. de Pina Filho AC (ed) (2011) Biped robots. In Tech, Vienna (Open Access)
6. Webb B, Consi TR (2001) Biorobotics. MIT Press, AAI Press, Cambridge, MA
7. Lozano R (ed) (2010) Unmanned aerial vehicles: embedded control. Wiley, Hoboken, NJ
8. Roberts GN, Sutton R (2006) Advances in unmanned marine vehicles. IET Publications, London, UK
9. Saridis G (1985) Advances in automation and robotics. JAI Press, Greenwich
10. Tzafestas SG (1991) Intelligent robotic systems. Marcel Dekker, New York

11. Antsaklis PJ, Passino KM (1993) An introduction to intelligent and autonomous control. Kluwer, Springer, Norwell, MA
12. Jacak W (1999) Intelligent Robotic systems: design, planning and control. Kluwer, Plenum, New York, Boston
13. Nof S (1999) Handbook of industrial robotics. Wiley, New York
14. Speich JE, Rosen J (2004) Medical robotics, encyclopedia of biomaterials and biomedical engineering 983–993
15. Da Vinci Surgical System. http://intuitivesurgical.com
16. Schraft RD, Schmierer G (2000) Service robots. Peter AK, CRC Press, London
17. Katevas N (2001) Mobile robotics in healthcare. IOS Press, Amsterdam, Oxford
18. Cook AM, Hussey SM (2002) Assistive technologies: principles and practice, St. Luis, Mosby
19. Tzafestas SG (ed) (1998) Autonomous mobile robots in health care services (special issue). J Intell Robot Syst **22**(3–4):177–374
20. Sheridan TB (1992) Telerobotics, automation and human supervisory control. MIT Press, Cambridge, MA
21. Moray N, Ferrell WR, Rouse WB (1990) Robotics control and society. Taylor & Francis, London
22. Votaw B. Telerobotic applications. http://www1.pacific.edu/eng/research/cvrg/members/bvotaw/
23. White SD (2007) Military robots. Book Works, LLC, New York
24. Zaloga S (2008) Unmanned aerial vehicles: robotic air warfare 1917–2007. Osprey Publishing, Oxford
25. Curtiss ET, Austis E (2008) Educational and entertainment robot market strategy, market shares, and market forecasts 2008–2014. Winter Green Research, Inc, Lexington, MA
26. Fong T, Nourbakhsh IR, Dautenhahn K (2003) A survey of socially interactive robots. Robot Auton Syst 42(3–4):143–166
27. Dautenhahn K (2007) Socially intelligent robots: dimensions of human-robot interaction. Philos Trans-Royal Soc London Biol Sci B 362:679–704
28. Dautenhahn K, Robins B, Wearne J (1995) Kaspar: kinesis and synchronization in personal assistant robotics. Adaptive Research Group, University of Hertfordshire
29. Breazal CL (2002) Designing sociable robots. MIT Press, Cambridge, MA
30. Breazeal CL (2000) Sociable machines: expressive Social Exchange between Humans and Robots. Ph.D. Thesis, Department of Electrical Engineering and Computer Science, Cambridge, MA
31. Stanton CM, Kahn PH Jr, Severson RL, Ruckert JH, Gill BT (2008) Robotic animals might aid in the social development of children with autism. In: Proceedings of 3rd ACM/IEEE international conference on human robot interaction. ACM Press, New York

Chapter 5
Roboethics: A Branch of Applied Ethics

You realize, there is no free-will in anything we create with artificial intelligence

Clyd Dsouza

The bottom line is, robots need to be responsive and resilient. They have to be able to protect themselves and also smoothly transfer control to human when necessary.
("Want responsible robots? Start with responsible humans")

David Woods

5.1 Introduction

As we have discussed in Chap. 2, *applied ethics* is the branch of ethics which typically starts with an ethical theory (i.e., a set of moral/ethical guides) and then applies it to particular areas of human life and society in order to address specific ethical dilemmas arising therein. Likewise *roboethics* is the branch of applied ethics which is concerned with addressing the following three issues [1]:

- The ethics of people who create and employ robots.
- The ethical system embedded into robots.
- The ethics of how people treat robots.

Questions that have to be considered in the framework of roboethics include:

- What role would robots have into our future?
- Is it possible to embed into robots ethics codes, and if yes is it ethical to program robots to follow such codes?
- Who or what is responsible if a robot causes harm?
- Are there any types of robot that should not be designed? Why?
- How might human ethics be extended such that to be applicable to the combined human-robot actions?
- Are there risks in creating emotional bonds with robots?

S.G. Tzafestas, *Roboethics*, Intelligent Systems, Control and Automation: Science and Engineering 79, DOI 10.1007/978-3-319-21714-7_5

The three principal positions of robotics scientists about roboethics are [2]:

- **Not interested in roboethics** These scientists argue that the action of robot designers is purely technical and does not have a moral or social responsibility in their work.
- **Interested in short-term robot ethical issues** This is the attitude of those who consider the ethical performance in terms of *good* or *bad*, and adopt certain social or ethical values.
- **Interested in long-term robot ethical issues** Roboticists having this attitude express their robotic ethical concern in terms of global, long term aspects.

In general, the technological and computer advancements are continuing to promote reliance on automation and robotics, and autonomous systems and robots live with people. It therefore follows that the ethical examination of robot creation and use makes sense both in *short-term* and *long-term*. Roboethics is a human centered ethics and so it must be compatible with the legal and ethical principles adopted by the international human rights organizations.

The purpose of this chapter is to outline a set of general fundamental issues of roboethics addressed by robotics scientists over the years.

Specifically, the chapter:

- Provides a preliminary general discussion of roboethics.
- Presents the top-down roboethics approach (deontological roboethics, consequentialist roboethics).
- Provides a discussion of the bottom-up roboethics approach.
- Outlines the fundamental requirements for smooth human-robot symbiosis.
- Addresses a number of questions related to the robot rights issue.

5.2 General Discussion of Roboethics

Robotics on its own is a very sensitive field since robots are closer to humans than computers (or any other machines) that may ever created, both morphologically and literally [3]. This is because of their shape and form, which remind us of ourselves. Robots must not be studied on their own, separated from sociotechnical considerations of today's societies. Scientists should have in mind that robots (and other high-tech artifacts) may influence how societies develop in ways that could not be anticipated during their design. A dominant component of human concern about robotics is *roboethics*. The more *autonomy* is provided and allowed to a robot the more moral and ethical sensitivity is required [4]. Currently, there is no special legislation about robots and particularly about cognitive capable (intelligent) robots. Legally, robots are treated in the same way as any other technological equipment and artifact. This is probably due to the fact that robots with full intelligence and autonomy are not yet in operation or in the market. However, many roboticists have the opinion that even with the present pace of advancement of artificial intelligence and robotic engineering such laws will be required soon.

As mentioned in the introduction, Asaro in [1], argues that there are three distinct aspects in roboethics:

- How to design robots to act ethically.
- How humans must act ethically taking on their shoulders the ethical responsibility.
- Theoretically, can robots be fully ethical agents?

All these aspects must be addressed in a desirable roboethics framework. This is due to that these three aspects represent different faces of how moral responsibility should be distributed in socio-technical frameworks involving robots, and people and robots ought to be regulated.

The primary requirement from a robot (and other autonomous agents) is not doing harm. The issue of resolving the vague moral status of moral agents and human ethical dilemmas or ethical theories is a must but at a secondary level. This is because as robots get more capabilities (and complexity) it will become necessary to develop more advanced safety control measures and systems that prevent the most critical dangers and potential harms. Here, it should be remarked that the dangers for robots are not different from those of other artifacts in our society from factories, to internet, advertising, political systems, and weapons. As seen in Chap. 4, robots are created for a purpose, i.e., for performing tasks for us and release people from various heavy or dangerous labors. But they do not replace us or eliminate our need or desire to live our lives.

In many cases, accidents in the factory or in society are attributed to faulty mechanisms, but we rarely ascribe moral responsibility to them. Here, we mention the National Rifle Association slogan: “Guns don’t kill people, people kill people is only partially correct. Actually, it is “people + guns that kill people” [5]. According to this point of view, in a car accident it is the “human driver + car” agent responsible for the accident.

In roboethics it is of great importance to recognize fundamental differences between human and computer or robot intelligence; otherwise we might arrive at mistaken conclusions. For instance, on the basis of our belief that human intelligence is the only intelligence and our desire is to get power, we tend to assume that any other intelligent system desires power. But for example, though *Deep Blue* can “win” at chess World Chess Champions, it has absolutely no representation of power, or human society anywhere in its program [6].

At present fully intelligent robots do not exist, since for a complete artificial intelligence there are many obstacles that must be overcome. Two dominant obstacles of this kind are *cognition* and *creativity*. This is because there are still no comprehensive models of cognition and creativity. The boundaries of living/nonliving and conscious/unconscious categories are not yet well established and have created strong literary and philosophical debates. But even a satisfactory establishment of those boundaries would exist there is no certainty that the answer to the ethical questions would be made easier [1].

Naturally, different cultures have different views on autonomy, dignity, and morality. As a consequence, possible "*ethical robots*" of the future that may be developed in societies with different cultures would have embedded different ethical codes, something which complicates the transfer of robots from one culture to another. But despite the differences in moral codes from culture to culture, the codes have much in common. This common code part is called "*moral deep structure*". Issues on "multicultural roboethics" are discussed in [7] and in Chap. 10 of this book.

Over the years a large number of papers and books have been published dealing with particular roboethics questions adopting different ethical views. It is noted that like the fact that it is difficult to define universally acceptable ethical principles, it seems very difficult (if not impossible) to homogenize all these attempts.

As Wallach explains, the implementation of robot (machine) ethics has two basic approaches, namely:

- *Top-down approach*: In this approach, a certain number of ethical rules or principles governing moral behavior are prescribed and embodied in the robot system.
- *Bottom-up approach*: *Here*, an evolutionary or developmental psychological-like way is followed to learn appropriate responses to moral considerations. This is analogous to how growing children learn "morality" (what is right and wrong) based on social context and experience.

5.3 Top-Down Roboethics Approach

The *top-down* approach can be applied to both deontological or consequentialist/utilitarian theories.

5.3.1 Deontological Roboethics

Deontological approaches to ethics, such as Kant's *categorical imperative,* utilitarianism, the Ten Commandments or Isaak Asimov's Laws, provide high flexibility, but because they are too broad or abstract, they may be less applicable to specific situations. As we have seen in Chap. 2, in a deontological theory, actions are evaluated for their own rather than by the consequences or the utility value they produce. Actions implement "*moral duties*" and can be considered as innately right (or wrong) independently of the actual consequences they may cause.

Regarding robotics the first ethical system proposed is that of the three *Asimov's laws* first appeared together in his story "*Runaround*" [8].

These three laws are:

Law 1: A robot may not injure a human being or, through inaction, allow a human being to come to harm.
Law 2: A robot must obey orders it receives from human beings, except when such orders conflict with the first law.
Law 3: A robot must protect its own existence as long as such protection does not conflict with the first or second laws.
Later Asimov added a law which he named "*Law Zero*", since it has a higher importance than laws 1 through 3. This law states:
Law 0: No robot may harm humanity or through inaction allow humanity to come to harm.

These laws are *human-centered* (anthropocentric), i.e., they consider the role of robots in human service, and imply that robots have sufficient intelligence (perception, cognition) to make moral decisions following the rules in all situations how matter their complexity.

In [9] these rules are formulated such as to support logical reasoning. This is done using a suitable classification scheme of ethical actions which simplifies the process of determining which robotic action is the most ethical in complicated cases. Therefore, given the current maturity level of intelligent robots these laws, despite their superior elegancy and simplicity, cannot at present provide a practical basis for roboethics. However, although Asimov's laws are still fictional, they seem to occupy an important position in present-day ethical, legal and policy considerations of the activity, regard and use of robots.

A modern *deontological ethical* system involving ten rules was proposed in [10, 11]. These rules are:

1 Don't kill	6. Don't deceive
2 Don't cause pain	7. Keep your promise
3 Don't disable	8. Don't cheat
4 Don't deprive of freedom	9. Obey the law
5 Don't deprive of pleasure	10. Do your duty

This is a multi-rule ethical system, and as in all multi-rule systems, it is possible to face a conflict between the rules. To address the rule conflict problem one may treat the ethical rules as *dictating prima facie duties* [12]. This, for example, means that if an agent gives a promise, it has the obligation to keep the promise. For other things that are equal, this promise should also be kept for them. Rules may have exceptions, and moral considerations derived from other rules, may override the rule. As argued in [10] these rules are not absolute. A way for deciding when it is okay not to follow a rule is provided in [10]. This rule is the following:

"Everyone is always to obey the rule except when an impartial rational person can advocate that violating it be publicly allowed. Anyone who violates the rule, when an impartial rational person could not advocate that such a violation may by publicly allowable, may be punished".

Aquinas natural law-based virtue system involves the following virtues: faith, hope, love, prudence, fortitude, temperance and justice. The first three are "theological virtues", and the other four "human virtues". Therefore in rule (deontological) form this system is:

- Act with faith
- Act with hope
- Act with love
- Act prudently
- Act with fortitude
- Act temperately
- Act justly

The same is true for all virtue ethical systems (Aristotle, Plato, Kant, etc.). All these systems can be implemented in deontological rule-based system form.

In [12, 13] it is argued that for a robot to be ethically correct the following conditions (*desiderata*) must be satisfied:

> "$\mathbf{D_1}$: Robots only take permissible actions.
> $\mathbf{D_2}$: All relevant actions that are obligatory for robots are actually performed by them, subject to ties and conflicts among available actions.
> $\mathbf{D_3}$: All permissible (or obligatory or forbidden) actions can be *proved* by the robot (and in some cases, associated systems, e.g., oversight systems) to be permissible (or obligatory or forbidden), and all such proofs can be explained in ordinary English".

The above ethical system can be implemented in top-down fashions. The following four top-down approaches are discussed in [13]:

- **Approach 1**: *Direct formalization and implementation of an ethical code under an ethical theory using deontic logic* [14].

Standard *deontic logic* **(SDL)** has two inference rules and three axioms [13]. SDL has many useful features, but it does not formalize the concepts of actions being obligatory (or permissible or forbidden) for an agent. In [15] an AI-friendly semantics has been proposed which, using the axiomatizations studied in [16], has regulated the behavior of two robots in an ethically sensitive case study using deontic logic.

- **Approach 2**: *Category theoretic approach to robot ethics.*

This theoretic approach is a very useful formalism and has been applied to many areas ranging from set-theory-based foundations of mathematics [17] to functional programming languages [18]. In [19] the robot *PERI* was designed which is able to make ethical correct decisions using reasoning from different logical systems viewing them from a category-theoretic perspective.

- **Approach 3**: *Principlism.*

In this approach the *prima facie duties* theory (*Ross*) is applied [20]. The three duties considered in medical ethics are:

- Autonomy
- Beneficence
- Nonmaleficence

Autonomy is interpreted as "allowing patients to make their own treatment decisions". *Beneficence* is interpreted as "improving patient health", and *nonmaleficence* is interpreted as "doing no harm". This approach was implemented in the advising system **Med Eth Ex**, which via computational inductive logic infers sets of consistent ethical rules from the judgments made by bioethics.

- **Approach 4**: *Rules of Engagement.*

In [21], a comprehensive architecture was proposed for the ethical regulation of autonomous robots that have destructive power. Using deontic logic and, among the elements of this architecture, specific military rules of engagement for what is permissible for the robot, a computational framework was developed. These rules of engagement are referred to as the "*ethical code for controlling a lethal robot*". The rules may be dictated by some society or nation, or they may have a utilitarian nature, or, entirely differently, could be viewed by the human as coming directly from God. In [13] it is argued that such a top-down deontologic code though it is not widely known, it provides a very rigorous approach to ethics which is known as "*divine command ethics*" [22].

5.3.2 Consequentialist Roboethics

In *consequetialist* theory the morality of an action is judged by its consequences. The best present moral action is the action that leads to the best future consequences. In utilitarian theory the best future consequences are determined or predicted using a certain *goodness measure*. The drawbacks of this approach have been discussed in Sect. 2.3.3. Actually, utilitarianism uses a mathematical framework for determining the best action choice by computing and maximizing "*goodness*", however defined, for all actions.

The basic requirements for a robot to be able to reason and act along the consequentialistic/utilitarian ethical theory are the following:

- To be able to describe every situation in the world.
- To be able to produce alternative actions.
- To be able to predict the situation(s) that would be the consequence of taking an action given the present situation.
- To be able to evaluate a situation in terms of its goodness or utility.

These requirements do not necessarily mean that the robot must have high-level artificial intelligence features, but high-level computational ones. The ethical correctness of an action is determined by the "*goodness*" criterion selected for evaluating situations. Actually, many evaluation criteria were proposed over the years which were formed such that to balance pleasure over pain for all persons in the society in an "*aggregate*" way. Specifically, let m_{p_i} be the measure of pleasure (or goodness) for the person i, and h_i the weight assigned to each person. Then the utility criterion function to be maximized has the general form:

$$J = \sum_i h_i m_{p_i}$$

where i extends over all persons of the population.

- In ideal (*universalist*) utilitarian approach the weights for all persons are equal, i.e., each person counts equally.
- In an *egoist* approach, the weight for the egoist person is 1, and for all other persons is zero.
- In an *altruist* approach, the weight for the altruist is zero and positive for all other persons.

The fundamental, commonly accepted, objection to utilitarianism is that *it is not necessarily just*. Although utilitarianism values the benefits brought to society as a whole, it does not guarantee that fundamental human rights and goodness of each individual [21] will be respected, and so it is of limited acceptability [23]. One way to address this problem of justice is to assign higher weight values to persons that are presently less happy or well-being, i.e., the well-being of the *less fortunate person* should count more than that of the more fortunate. In many statistical studies it was verified that only a few people conform to the utilitarian ideal ($h_i \equiv h_j$ for all i, j). For example, most people consider as more important their relatives or people they know better, i.e., they give greater h_i value to their relatives or to people they know better. The method for weight selection depends on the agent's value theory or axiology. The basic issue here is: "what exactly is the m_{p_i} measure?" In *hedonistic* attitude it is believed that the good is pleasure and the bad is pain, while in other cases the ethical aim is to maximize happiness.

5.4 Bottom-Up Roboethics Approach

In this approach the robots are equipped with computational and AI capabilities to adapt themselves in some way to different contexts, such that to be able to act properly in complicated situations. In other words the robot becomes able to learn, starting from perception of the world using a set of sensors, proceeds further to the planning of actions based on the sensory data, and then finally executes the action

[24]. Very often, the robot is not going directly to the execution of the decided action, but via intermediate corrections. This process is similar to the way children learn their ethical performance from their parents through teaching, explanation and reinforcement of good actions. In overall, this kind of moral learning falls within the *trial-and-error* framework. A robot which learns in this child-like way has been developed at MIT, named "*Cog*". The learning data for Cog are acquired from the surrounding people [25, 26]. The learning tool used is the so-called *neural network learning* which has a subsymbolic nature, in the sense that instead of clear and distinct symbols, a matrix of synaptic weights is used that cannot be interpreted directly [27]. It should be emphasized that when the neural network learning (weights) is implemented in new situations it is not possible to predict accurately the robot's actions. This means, in some way, that robot manufacturers are no longer the only responsible for the actions of the robot. The responsibility is distributed between the robot manufacturer (the robotics expert who designed and implemented the learning algorithm) and the robot owner (user) who is not an expert in robotics. Here we have the ethical issue that in all cases (even with learning robots) the human role as the decision maker for the man-robot interaction must be assured, and the legal issue that responsibility is divided between the robot's owner and its manufacturer. A comprehensive discussion of bottom-up and top-down approaches to roboethics is provided in [4, 28]. It is argued there that an ethical learning robot needs both top-down and bottom-up approaches (i.e., a suitable hybrid approach). Some of the ethical rules are embodied in a top-down mode, while others are learned in a bottom-up mode. Obviously, the hybrid approach is more powerful since the top-down principles are used as an overall guide, while the system has the flexibility and moral adaptability of the bottom-up approach.

In [4] it is argued that the morality of robots is distinguished in:

- Operational morality
- Functional morality
- Full morality

In *operational morality* the moral significance and responsibility lies totally in the humans involved in their design and use, far from full moral agency. The computer and robot scientists and engineers designing present day robots and software can generally forecast all the possible situations the robot will face.

Functional morality refers to an ethical robot's ability to make moral judgments when deciding a course of action without direct top-down instructions from humans. In this case the designers can no longer predict the robot's actions and their consequences.

Full morality refers to a robot which is so intelligent that it is entirely autonomously selecting its actions and so it is fully responsible for them. Actually, moral decision making can be regarded as a natural extension to engineering safety for systems with more intelligence and autonomy.

5.5 Ethics in Human-Robot Symbiosis

The primary objective of *human-robot symbiosis* (living together, partnership) is to fill in the gap between full autonomous and human-controlled robots. Robotic systems must incorporate the human's needs and preferences in all cases. In practice, a dynamic subdivision of the work of the human and the robot must be made such that to optimize the admissible task range, accuracy and work efficiency of the system (*shared autonomy*). To this end, several fundamental technical issues should be addressed, namely [29]:

- Human-robot communication.
- Human-robot architecture.
- Autonomous machine learning (via observation and experience).
- Autonomous task planning.
- Autonomous execution monitoring.

The human-robot system must be regarded as a *multi-agent system* where the human-robot interaction is split in:

- *Physical part* which refers to the structure of the body of the human and the robot.
- *Sensory part* which refers to the channels via which the human and the robot take information about each other and the world.
- *Cognitive part* which refers to the issues of internal function of the system. For the human this includes the mind and affective state. For the robot it includes the reasoning mode and the capability to communicate the intention.

The human agent is supported by a number of specialized agents. The primary agents should include the following:

- *Monitoring agent* which monitors passively human features (e.g., physical features and emotional state). This is implemented via a number of technologies (e.g., speech recognition, sound localization, motion detection, face recognition, etc.).
- *Interaction agent* which has the ability to handle more pro-active functions of the interaction, such as handling communication via an interaction agent or modeling the interaction with the human via the ethical/social agent.
- *Ethical social agent* which contains a set of ethical and social rules of interaction that enable the robot to execute ethical actions and interact with people according to the accepted overall ethical/social context.

In a symbiotic system humans and robots cooperate in the decision-making and control tasks in complex dynamic environments in order to achieve a common goal. Humans must not be viewed as components of the symbiotic system in the same way as the robots or the computers. Human-machine/robot systems have a long history. For example, in [30] a vision of human-machine symbiosis is formulated as:

> "Men will set the goals, formulate the hypotheses, determine the criteria, and perform the evaluations. Computing machines will do the routine work that must be done to prepare the way for insights and decisions in technical and scientific thinking". Preliminary analysis indicates that symbiotic partnership will perform intellectual operations much more efficiently than man alone can perform them.

In [31] a number of important questions about human-robot symbiosis are formulated, namely:

- What are the consequences of the fact that today information technology devices are developed by computer scientists and engineers only?
- What is the meaning of the master–slave relation with regard to robots?
- What is the meaning of robot as a partner in different settings?
- How building social robots shapes our self-understanding, and how these robots impact our society?

These and other questions have received intensive attention over the years and many different answers already exist with many new ones to come.

5.6 Robot Rights

Current Western legislation considers that robots are a form of inanimate agent without duties or rights. Robots and computers are not legal persons and have no standing in the juridical system. It follows that robots and computers may not be the perpetrators of a crime. A human that dies in the arms of a robot has not been murdered. But what should happen if a robot has partial self-awareness and sense of self-preservation, and makes moral decisions? Should such moral robots possess the rights and duties of humans or some other rights and duties? This question has produced much discussion and debate among scientists, sociologists, and philosophers.

Many robotics scientists envision that after some decades, robots will be sentient and so they will need protection. They argue that a "*Bill of Rights for Robots*" must be developed. Humans need to exercise empathy to the intelligent robots they create, and robots must be programmed with a sense for empathy for humans (their creators). These people believe that greater empathy results in lower tendency of humans and robots to act violently and do harm. As stated in [32], "the hope of the future is not technology alone, it is the empathy necessary for all of us, human and robot, to survive and thrive".

The hard question for recognizing robots that are conscious and have feelings as beings with moral standing and interests, and not as objects of property, is the following:

"How could we be able to recognize that the robot is truly conscious and is not purely "mimicking" consciousness by the way it was programmed? If the robot simply mimics consciousness there is no reason to recognize moral or legal rights to it". But, it is argued in [33], that "if a robot is built in future with humanlike capabilities that might include consciousness there would be no reason to think that the robot has no real consciousness". This may be regarded as the starting point of assigning legal rights to robots [33].

In [34], the robot rights issue is discussed by exploring the human tendency to anthropomorphize social robots. This tendency increases when animals show behavior which is more readily associated with human consciousness and emotions. It is argued that since social robots are specifically designed to elicit anthropomorphic characteristics and capabilities and in practice humans interact with social robots differently than they do with other artifacts, certain types of protection of them would fit into our current legislation, particularly as analog to *animal abuse laws*. Another argument in the discussion of the possibility of extending legal protection to robotic companions/socialized robots, expanded in [34], is based on the approach that regards the purpose of law as a *social contract*. Laws are designed and used to govern behavior for the greater good of society, i.e., laws must be used to influence people's preferences, rather than the opposite. This suggests that costs and benefits to society as a whole must be evaluated in a utilitarian way. If the purpose of law is to reflect social norms and preferences, the societal desire for robot rights (if it exists) should be taken into account and converted to law [34]. Based on the Kantian philosophical argument for protecting animals, it is logically reasonable to extend this protection to *socialized robots*. But a practical difficulty to do this is to define the concept of "socialized robot" in a legal way. In overall, the question of whether socialized robots/robot companions (such as those presented in Sect. 8.4) should be legally protected is very complicated.

At the other end, many robotics researchers and other scientists argue strongly against giving to robots moral or legal responsibility, or legal rights. They state that robots are fully owned by us, and the potential of robotics should be interpreted as the potential to extend our abilities and to address our goals [35–37]. In [35], the focus is on the ethics of building and using "robotic companions", and the thesis is that "robots should be built, marketed and considered legally as slaves, not as companion peers". The interpretation of the statement: "*robots should be slaves*" is by no means that "*robots should be people you own*", but "*robots should be servants you own*".

The primary claims made in [35] are:

- It is good and useful to have servants, provided no one is dehumanized.
- A robot can be a servant without being a person.
- It is right and natural for people to own robots.
- It is not right to let people believe that their robots are persons.

The reasoning presented in [35] is that robots are completely our responsibility because, actually, we are the designers, manufacturers, owners, and users of them. Their goals and behavior are determined by us either *directly* (by specifying their intelligence), or *indirectly* (by specifying how they get their intelligence). Robot owners should not have any ethical obligation to robots that are their sole property beyond society's common sense and decency, which holds for any artifact. In conclusion, the thesis which is presented in [35] is: "Robots are tools, and like any other artifact when it comes in the domain of ethics. An autonomous robot definitely incorporates its own motivational structure and decision mechanisms, but *we* choose those motivations and design the decision-making system. All their goals are derived from us … Therefore, we are not obliged to the robots, but to the society".

5.7 Concluding Remarks

In this chapter we have discussed the basic issues of roboethics considering the robots as sociotechnical agents. Roboethics is closely related to robot autonomy. The ethical issues arising come from the robot progress and achievement of more autonomy or more cognitive features. Asimov's laws are anthropocentric and assume tacitly that robots can get sufficient intelligence such as to be able to make correct moral decisions under all conditions. These laws were used by several authors as a basis for developing and proposing more realistic laws of a deontologic nature. These laws as well as rules of consequentialist nature are embodied into the robot's computer in a top-down approach. The alternative way proposed for embodying ethical performance into robots is the bottom-up ethical learning approach (e.g., via neural learning or other learning schemes). Another issue considered in the chapter is the ethical human-robot symbiosis. Human-robot integration goes beyond the level of the single individual and addresses the issue about how society could and should look at a human-robot society. Naturally, this includes aspects of human and robot rights. Of course, the question of robot rights is an issue of strong debate. Here, it is mentioned that Japan and Korea have started developing policies and laws to guide and govern human-robot interactions. Motivated by Asimov's laws Japanese Government has issued a set of bureaucratic provisions for "logging and communicating any injuries robots cause to humans in a central data base". Korea has developed a code of ethics for human-robot interaction (*Robot Ethics Charta*) which defines ethical standards that would be programmed into robots, and limit some potential abuses of robots by humans [38] (see Sect. 10.6). Actually, no single generally accepted moral theory exists, and just a few generally accepted moral norms exist. On the other hand, although a multiple legal interpretation of cases exists, and judges have different opinions, the legislation system seems to offer a safer framework and tends to do a pretty good job in addressing issues of responsibility in both civil law and criminal law. So, starting to think from a legal responsibility perspective it is more likely to arrive at correct practical answers [1].

References

1. Asaro PM (2006) What should we want from a robot ethics? IRIE Int Rev Inf Ethics 6(12):9–16
2. Verrugio G, Operto F (2006) Roboethics: a bottom-up interdisciplinary discourse in the field of applied ethics in robotics. IRIE Int Rev Inf Ethics 6(12):2–8
3. Lichocki P, Kahn PH Jr, Billard A (2011) A survey of the robotics ethical landscape. IEEE Robot Autom Mag 18(1):39–50

4. Wallach W, Allen C (2009) Moral machines: teaching robots right from wrong. Oxford University Press, Oxford
5. Latour B (1999) Pandora's hope: essays on the reality of science studies. Harvard University Press, Cambridge
6. Hsu FH (2002) Behind deep blue: building the computer that defeated the world chess champion. Princeton University Press, Princeton
7. Wagner JJ, Van der Loos HFM (2005) Cross-cultural considerations in establishing roboethics for neuro-robot applications. In: Proceedings of 9th IEEE international conference on rehabilitation robotics (ICOOR'05), Chicago, LL, 28 June–1 July 2005, pp 1–6
8. Asimov I (1991) Runaround. Astounding science fiction, Mar 1942. Republished in Robot Visions. Penguin, New York
9. Al-Fedaghi SS (2008) Typification-based ethics for artificial agents. In: Proceedings of 2nd IEEE international conference on digital ecosystems and technologies (DEST'08), Phitsanulok, Thailand, pp 482–491
10. Gert B (1988) Morality. Oxford University Press, Oxford
11. Gips J (1992) Toward the ethical robot. In: Ford K, Glymour C, Mayes P (eds) Android epistemology. MIT Press, Cambridge. http://www.cs.bc.edu/~gips/EthicalRobot.pdf
12. Bringsjord S (2008) Ethical robots: the future can heed us. AI Soc 22(4):539–550
13. Brinsjord S, Taylor J (2011) The divine command approach to robotic ethics. In: Lin P, Abney K, Bekey GA (eds) Robot ethics: the ethical and social implications of robotics. The MIT Press, Cambridge
14. Aqvist E (1984) Deontic logic. In: Gabbay D, Guenthner F (eds) Handbook of philosophical logic. Extensions of classical logic, vol II. D. Reidel, Dordrecht
15. Horty J (2001) Agency and deontic logic. Oxford University Press, New York
16. Bringsjord S, Arkoudas K, Bello P (2006) Towards a general logistic methodology for engineering ethically correct robots. IEEE Intell Syst 21(4):38–44
17. Marquis J (1995) Category theory and the foundations of mathematics. Synthese 103:421–427
18. Barr M, Wells C (1990) Category theory for computing science. Prentice Hall, Upper Saddle River
19. Brinsjord S, Taylor J, Housten T, van Heuveln B, Clark M, Wojtowicz R (2009) Plagetian roboethics via category theory: moving beyond mere formal operations to engineer robots whose decisions are guaranteed to be ethically correct. In: Proceedings of ICRA-09 workshop on roboethics, Kobe, Japan, 17 May 2009
20. Anderson M, Anderson S (2008) Ethical health care agents. In: Sordo M, Vaidya S, Jain LC (eds) Advanced computational intelligence paradigms in healthcare. Springer, Berlin
21. Arkin R (2009) Governing lethal behavior in autonomous robots. Chapman and Hall, New York
22. Quinn P (1978) Divine commands and moral requirements. Oxford University Press, New York
23. Grau C (2006) There is no "I" in "robot": robots and utilitarianism. IEEE Intell Syst 21(4):52–55
24. Decker M (2007) Can humans be replaced by autonomous robots? Ethical reflections in the framework of an interdisciplinary technology assessment. In: Proceedings of IEEE international conference on robotics and automation, Rome, Italy, 10–14 Apr 2007
25. Brooks RA (1997) The Cog project. J Robot Soc Jpn 15(7):968–970
26. Brooks RA (1994) Building brain bodies. Auton Robots 1(1):7–25
27. Matthias A (2004) The responsibility gap: ascribing responsibility for the actions of learning automata. Ethics Inf Technol 1:175–183
28. Wallah W, Allen C, Smit I (2007) Machine morality : bottom-up and top-down approaches for modeling moral faculties. AI Soc 22(4):565–582
29. Kawamura K, Rogers TE, Hambuchen K, Erol D (2003) Toward a human-robot symbiotic system. Robot Comput Integr Manuf 9:555–565
30. Licklider JRC (1960) Man-computer symbiosis. IRE Trans Hum Factors Electron HFE-1: 4–11
31. Capurro R (2009) Ethics and robotics. In: Capurro R, Nagenborg M (eds) Ethics and robotics. Akademische Verlagsgesellschaft, Heidelberg, pp 117–123

32. Moore KJ (2011) A dream of robot's rights. http://hplusmagazine.com/2011/08/29/a-dream-of-robots-rights
33. Singer P, Sagan A (2009) Do humanoid robots deserve to have rights? The Japan Times, 17 Dec 2009. www.japantimes.co.jp/opinion/2009/12/17/commentary/do-humanoid-robots-deserve-to-have-rights
34. Darling K (2012) Extending legal rights to social robots. http://ssrn.com/abstract=2044797
35. Bryson JJ (2008) Robots should be slaves. In: Wilks Y (ed) Close engagements with artificial companions: key social, psychological, ethical and design issue. John Benjamins, Amsterdam
36. Bryson JJ, Kime P (1998) Just another artifact: ethics and the empirical experience of AI. In: Proceedings of 15th international congress on cybernetics, Namur, Belgium, pp 385–390
37. Bryson JJ (2000) A proposal for the Humanoid Agent-builders League (HAL). In: Barnden J (ed) Proceedings of AISB'00 symposium on artificial intelligence, ethics and (Quasi-) human rights, Birmingham, U.K., pp 1–6, 17–20 Apr 2000
38. Lovgren S (2007) Robot code of ethics to prevent android abuse and protect humans. National Geographic News, 16 May 2007. http://news.nationalgeographic.com/news/2007/03/070316-robot-ethics.html

Chapter 6
Medical Roboethics

Our expectations for a technology rise with its advancement.

Henry Petroski

In law a man is guilty when he violates the rights of others In ethics he is guilty if he only thinks of doing so.

Immanuel Kant

6.1 Introduction

Medical roboethics (or *health care roboethics*) combines the ethical principles of medical ethics and roboethics. The dominant branch of medical robotics is the field of robotic "*surgery*" which is receiving an increasingly stronger position in modern surgery. The proponents of robotic surgery advocate that robots assist surgeons to perform surgery with enhanced access, visibility and precision with overall result the reduction of pain and blood loss, reduction of hospital stays, and finally allowing the patients to go back on normal life more quickly. However, there are many scientists and medical professionals that argue against this. For example in a study carried-out in an American Medical School (2011) it is concluded that there is a "lack of evidence that robotic surgery is any better, or any more effective, than conventional operations". Robotic surgery is very costly. Therefore the question which immediately rises is: "When there is marginal benefit from using robots, is it ethical to impose financial burden on patients or medical system?"

Another important area of health care robotics is "*rehabilitation/assistive robotics*". The ethical issues of this field which deals with assisting, through robotics, persons with special needs and elderly people to increase their mobility and other physical capabilities will be discussed in the next chapter.

In the present chapter we will be concerned with the general topic of medical ethics and the particular topic of ethics in robotic surgery. Specifically the chapter:

S.G. Tzafestas, *Roboethics*, Intelligent Systems, Control and Automation: Science and Engineering 79, DOI 10.1007/978-3-319-21714-7_6

- Provides a short general discussion of medical ethics.
- Provides an outline of the basic issues of robotic surgery.
- Discusses the particular ethical issues of robotic surgery.

6.2 Medical Ethics

Medical ethics (or *biomedical ethics, or health care ethics*) is a branch of applied ethics referring to the fields of medicine and health care [1–7]. Medical ethics also includes the *nursing ethics*, which is sometimes regarded as a separate field. The initiation of medical ethics goes back to the work of *Hippocrates* who has formulated the well known *Hippocratic Oath*. This Oath (*Όρκος* = Orcos, in Greek) is the most widely known of Greek medical texts. It requires a new physician to swear upon a number of healing gods that he will uphold a number of professional ethical standards.[1] The Hippocratic Oath has been rewritten over the centuries, to fit the values of different cultures influenced by Greek medicine. Today there are modern versions that use the term *maxim* ("*do no harm*") standing for the classical requirement of the oath that "*physicians should keep patients from harm*". The modern version of the Oath for today's medical students (widely adopted) is that written in 1964 by *Dr. Louis Lasagna*, dean of the School of Medicine at Tufts University.

Later in 2002 a coalition of international foundations introduced the so-called *Charter on Medical Professionalism*, which calls the doctors to uphold the following three fundamental principles:

- **Patient welfare** A patient's health is *paramount*.
- **Patient autonomy** A doctor serves to advise a patient only on health care decisions, and a patient's own choices are *essential* in determining personal health.
- **Social Justice** The medical community works to *eliminate disparities* in resources and health care across regions, cultures and communities as well as to abolish discrimination in health care.

The charter involves a "*set of hierarchical professional responsibilities*" that are expected from doctors.

The central issue in medical ethics is that medicine and health care deal with human health, life, and death. Medical ethics is concerned with ethical norms for the practice of medicine and health care, or how it ought to be done. Therefore it is clear that concerns of medical ethics are among the most important and influential in human life.

[1]For full information a translation in English of the Hippocratic Oath is given in the Appendix (Sect. 6.5) [8].

Reasons that call for medical ethics include (not exhaustively) the following:

- The power of physicians over human life.
- The potential for physicians and associated care givers to misuse this power or to be careless with it.

Other ethical concerns in medical practice are:

- Who and how decides to keep people technically alive by hooking them up to various machines? In case of disagreement between the doctors and the patient or his/her family members, whose opinion is ethically correct to follow?
- In the area of organ (kidney, lung, heart, etc.) transplants, which of the patients needing them should get the available organs, and what are the criteria that should be employed for the decision?
- Is health care a positive human right such that every person who needs it should have equal access to the most expensive treatment regardless of ability to pay?
- Is society obliged to cover the medical care of the public at large, via taxation or whatever rate increase?
- Who is ethically obliged to cover the costs of hospital treatment of indigent patients? The hospital, the state, or the paying patients?

These and other hard questions have to be carefully addressed by medical ethicists, from all viewpoints.

Over the years, ethicists and philosophers have been concerned with medical ethics and attempted to provide principles for health care providers (doctors, nurses, physical therapists). Actually, their principles were based on the classical applied ethics principles, namely:

- *Consequentialism* (utilitarian, teleologic) theories.
- *Kantian* (deontological, non-consequentialism) theory.
- *Prima facie duties theory.*
- *Case-based (casuistry) theory.*

In overall, medical ethics is ethics appealing traditional practical moral principles such as:

- Keep promises and tell the truth (except when telling the truth results in obvious harm).
- Do not interfere with the lives of other people unless they ask for this sort of help.
- Do not be so selfish that the good of others in never taken into account.

The *six-part approach* to medical ethics or *Georgetown mentra* (from the home city of their proponents) suggests that all medical ethical decisions should include the following principles (which involve the principles of the *Charter of Medical Professionalism*):

- *Autonomy* (The patients have the right to accept or refuse their treatment).
- *Beneficence* (The doctor should act in the best interest of the patient).

- *Non-maleficence* (The practitioner should "first not to do harm").
- *Justice* (The distribution of scarce health resources and decision of who gets what treatment should be just.).
- *Truthfulness* (The patient should not be lied and deserves to know the whole truth).
- *Dignity* (The patient has the right to dignity).

The above principles are not exhaustive and do not by their own give the answers as to how to treat a particular situation, but may give practical guide to doctors how they ought ethically treat real situations. Actually, doctors should have in mind the union of the various ethical codes, which however sometimes are in contradiction and may lead to ethical dilemmas. One such dilemma occurs when a patience refuses life-saving treatments in which case there occurs a contradiction between autonomy and beneficence.

An authoritative code of medical ethics has been developed and released by the *American Medical Association* **(AMA)**. For full information, the principles of this code are given in the Appendix (Sect. 6.5). This code is accompanied by a set of opinions on social policy issues [7–9].

6.3 Robotic Surgery

Before proceeding to the discussion of various ethical issues of *robotic surgery* it is useful to outline first the robotic surgery procedure and the resulting advantages over conventional surgery. Robotic surgery is a technique in which the surgeon performs surgery with the aid of a robot equipped with proper small tools. Robotic surgery is being applied for over two decades and involves robotic systems and image processing to interactively assist a surgeon in the planning and execution of the surgical procedures. Robotic surgery can be used for a number of different procedures, namely [10–15]:

- Radical prostatectomy
- Mitral valve repair
- Kidney transplant
- Coronary artery bypass
- Hip replacement
- Kidney removal
- Hysterectomy
- Gall bladder removal
- Pyloroplasty, etc.

Robotic surgery is not suitable for some complicated procedures (e.g., in certain types of heart surgery that require greater ability to move instruments in the patient's chest).

Actually, robotic surgery covers the entire operating procedure from the acquisition and processing of data to the surgery and post-operative examination. In the *preoperative phase*, the rigid body (such as bones) or deformable body (such as heart) of the patient are modeled in order to decide the targets of intervention. To this end, the particular features of medical imaging and the corresponding information are carefully examined and exploited. Then, the anatomic structures are used in order to schedule the operation's plan. The surgical tools (instruments) are inserted to the patient's body through small cuts, and under the surgeon's direction the robot matches the surgeon's hand movements to perform the procedure using the tiny instruments. A thin tube with a camera attached to the end of it (endoscope) allows the surgeon to view magnified 3-dimensional images of the patient's body on a monitor in real time. Robotic surgery is similar to laparoscopy surgery.

Typically, robotic-assisted laparoscopic surgery allows a less-invasive procedure that before was only possible with more invasive open surgery. The selected operation plan is correlated with the patient *intraoperative phase*. The robotic system assists in guiding the movement of the surgeon to achieve precision in the planned procedure. In many cases (such as hip replacement) the robot can work autonomously to carry out part of or the entire operating procedure.

The robotic-assisted surgical interventions contribute to enhancing the quality of care by minimizing the trauma (due to the reduction of the incision size, tissue deformity, etc.). Robotic-assisted surgery eliminates the effect of surgeon's hand tremor especially in procedures that last for several hours. In case of teleoperated robotic surgery, the surgeons work from a master console, and the surgeon's motions are filtered, reduced and transferred to the remote slave robot that performs the surgery on the body. With the advent of micro-robots the need for opening the patient will be eliminated.

In overall, the use of robots in the operating theatre enhances the surgeon's dexterity, precision/accuracy, and the patient's recovery time. Above all, as surgeons argue, it is expected that in the near future the robotic surgery will lead to the development of new surgical procedures that go beyond human capacity.

One of the first robots used in medical surgery was PUMA 560. Currently, there exist many surgery robots available commercially, such as the *Da Vinci* robot (Intuitive Surgical, Inc.), and *Zeus* robot (Computer Motion, Inc.) for "Minimally Invasive Surgery". The Da Vinci robot is shown in Fig. 4.11. For hip and knee replacement the *Acrobot* (Acrobot Company Ltd.) and *Caspar* (U.R.S.-Ortho GmbH) robots are marketed.

6.4 Ethical Issues of Robotic Surgery

Robotic surgery ethics, as a branch of medical ethics, includes at minimum the principles and guidelines discussed in the previous section, and the roboethics principles discussed in Chap. 5. Medical treatment (surgery and other) should first of all be legal. But a legal treatment may be not ethical. The legislation provides a baseline for people performance. Legislators and their laws are concerned with assuring that people behave in accordance to this minimum standard. The ethical standards are determined by the principles so far discussed, and in the context of licensed professionals (medical doctors, engineers, etc.) are provided by the accepted ethical code of each profession [16–19].

In the following, we discuss the legal part of injuring a patient in a robotic surgery. To this end, we briefly outline the basic rules of injury law. The law enforces on all individuals a duty of reasonable care to others, and determines it as how "*a reasonable (rational) person*" in the same situation would act. If a person causes injury to another, because of unreasonable action, then the law imposes liability on the unreasonable person. When the defendant is a surgeon (or any other professional who has a licence), the law will look at medical ethical rules as a guide.

That is, in a surgery malpractice suit, a plaintiff would try to establish that the surgeon's actions were at odds with the standards accepted by the medical community in order to prove that he breached his duty of care to the patient. Another essential law inquiry is "*causation*", for which the law requires proof of both "*actual*" and "*proximate*" causation before it imposes liability. Thus, in a lawsuit the plaintiff has to prove either that he would not have suffered injuries if it had not been for the defendant's actions (or in some cases that the defendant's actions were a "substantial factor" in bringing about the injury). To show *proximate (legal)* causation, a plaintiff has to prove that the defendant should have reasonable foreseen that his actions would cause the kind of injury that the plaintiff suffered. The above refers to the *legal liability in a personal injury*.

Now we will discuss the "*products' liability*". The manufacturer of a robot (or other device) has a duty to care the purchaser, and any other who might predict that he will come into contact with the robot. Thus, the manufacturer has the duty to design a safe robot and manufacture it free of failures, and to guarantee that the robot is fit for its normal purposes. This means that the manufacturer is liable for injuries caused from the robot's failure. If a doctor is using a robot that malfunctions and injure a patient (third party), the patient would like sue the doctor as the operator of the malfunctioning robot. For fair justice the law allows the doctor to request "contribution" from the robot manufacturer of the malfunctioned robot (i.e., a transfer of portion of the money he has to pay based on the fault of the robot manufacturer). If the doctor was entirely free of fault, then he will seek indemnification of the full damage by the manufacturer.

A dominant ethical issue related to robotic surgery is that of *social justice*. As discussed above, surgical robots are meant to improve the quality of life and dignity of the patient (reduction of patient's pain and recovering times, etc.). But this should go hand in hand with making such improvements available to all people without any discrimination. Unfortunately, very often high-tech medical treatment implies high costs mainly due to patenting rights (i.e., fees that have to be paid to the holders of the patient). The challenge here is to get and use high-tech medical aids but with affordable costs so as not to sharpen the differences between rich and poor. On this social justice issue the *European Group of Ethics* (**EGE**) has proposed as a practical moral solution the "*compulsory licence*" [20]. This should be the case whenever the access to medical diagnostics and treatment is blocked by misuses of patent rights. Clearly, the establishment of the legal procedure for the delivery of compulsory licence and the fair implementation in heath care is the duty of the state.

In the following we discuss a robotic surgery scenario which involves issues beyond the bounds of current personal injury law [19].

A patient with pangreatic tumor goes to Surgeon A, who is explaining to him the surgical procedure. The patient has provided *informed consent* for minimally invasive (laparoscopic) surgery, with the aid of a surgical robot (despite a number of risks involved in robotic surgery), and open surgery. The surgeon begins the surgery laparoscopically and finds that the tumor cannot be removed by conventional laparoscopic surgery. But by his experience he believes that the robot with its greater dexterity and accuracy can safely remove the tumor, which actually is the purpose of the robot. The surgeon A set ups and calibrates the robot and starts the operation of removing the tumor robotically when the robot malfunctions and injures the patient. The patient survives the operation but dies from cancer shortly after.

In case the patient's estate requests recovery of damages for the injuries due to surgery the following ethical issues arise:

- "*Was it ethical for the surgeon to offer the robotic surgery as an option for the patient, knowing the inherent risks?*"

To address this question, the code for medical ethics of the state where the surgery has taken place should be involved. For a U.S.A licenced surgeon and hospital, the American Medical Association code for medical ethics' opinion on *informed consent* states: "The physician's obligation is to present the medical facts accurately to the patient or to the individual responsible for the patient's care and to make recommendations for management in accordance to medical practice. The physician has an ethical obligation to help the patient make choices from among the therapeutic alternatives consistent with good medical practice". However a surgeon is not obliged to ask the patient whether he/she prefers him to use a certain surgical instrument or another. But under accepted standards in surgery it would be not ethical for the surgeon not to tell the patient about the use of the robot which differs very much from convention.

- "*Was it ethical for the surgeon to decide to use the robot?*"

This question is not different from the same question in other medical malpractice cases. To answer the question we have to look at what a reasonable surgeon would have done in the same situation. The legal question that has to be addressed is:

- "*Who should be legally liable for the patient's injury?*"

This case is complicated because of the death of the patient of his/her cancer. But, the patient would have died of cancer in the same way as he/she did, in fact, die after the robot was used.

Under the assumption that the patient's estate could sue for injuries incurred to the patient during the operation, the surgeon and the hospital would possibly seek indemnification from the robot manufacturer, sustaining that the robot's faulty behavior was the cause of the injury. Then, the manufacturer would likely sustain that the surgeon should not have opted to use the robot in this case and this, by doing so, the surgeon assumed the risk of injuring the patient.

Other legal and ethical issues related to the above scenario would be:

- The duty of the manufacturer to ensure that operators of the robot are adequately trained.
- The duty of the hospital to allow properly credential surgeons to use the robot.

A few words about robotic cardiac surgery would be useful to be given here. According to *Henry Louie* MD, "ironically, heart surgery has not really changed in approach since its inception from the late '70s. We still use needle and thread to construct bypass grafts, sew in or fix defective valves or close holes within the heart. Materials and imaging techniques to fix or see into the heart have dramatically improved however" [21]. In 2000, the FDA approved the use of the DaVinci robot for cardiac surgery, which using its tiny mechanical arms allows for a more accurate and less painful procedure, and decreases the patient's stay in the hospital by a couple of days. But, according to the doctors at Brown University, there is none to little difference when comparing heart surgery with the conventional method and the robotic method, and robotic surgery is slightly on the expensive side. Also the size of this robot does not fit the right criteria needed for cardiology, and the sheer size, especially in the area of pediatrics, presents a problem to cardiac surgery. *Dr. del Nido* (who does cardiac surgery in children) was also interviewed by Brown University and when asked about the major drawbacks of robotic surgery he replied that "the biggest drawback is that you don't have any sense of feel. You have no sensation. The robot gives you no tactile feedback. You are basically going by visual to what you are doing to the tissue and that is the biggest drawback". On the same issue of robotic heart surgery *Dr. Mark Grattan* M.D. in the question: "So how do you feel about the use of robotic surgery? Would you rather use that in future?", replied: "Not now, I don't think robotics has gotten to a point yet where it is safe to use on a lot of patients. To do robotics means that you have to do other things in order to protect the heart because you have to stop the heart" [21]. Some final legal and ethical issues of robotic surgery are the following:

- If robotic units are placed in different countries, what happens to the ethical and legislation considerations? For example, a doctor may have been licensed to perform robotic surgery in a particular location or jurisdiction. This may cause conflict if the operation itself is taking place in a jurisdiction other than the one he/she is licensed in.
- With the ability to perform telesurgery, doctors of richer countries are now able to monopolize the areas that previously, doctors from poorer countries would have occupied. This puts an additional gap between rich and poor countries, and gives them a large amount of power over those poorer countries.

- It is important to develop further the "*trustworthiness*" of robotic surgery. To this end, more randomized, controlled studies are needed. These are essential in truly determining if the fact that "robotic surgeries take more time, keep patients under anaesthesia longer and are more costly" is really worth the shortened recovery and minimally invasive incision? [22].

6.5 Appendix: Hippocratic Oath and American Medical Association Code of Ethics

6.5.1 Hippocratic Oath

"I swear by **Apollo** the physician, and **Asclepius**, and **Hygieia** and **Panacea** and all the gods and goddesses as my witness, that, according to my ability and judgment, I will keep this Oath and this contract:

- To hold him who taught me this art equally dear to me as my parents, to be a partner in life with him, and to fulfill his needs when required; to look upon his offspring as equals to my own siblings, and to teach them this art, if they shall wish to learn it, without fee or contract; and that by the set rules, lectures, and every other mode of instruction, I will impart a knowledge of the art to my own sons, and those of my teachers, and to students bound by this contract and having sworn this Oath to the law of medicine, but to no others.
- I will use those dietary regimens which will benefit my patients according to my greatest ability and judgment, and I will do no harm or injustice to them.
- I will not give a lethal drug to anyone if I am asked, nor will I advise such a plan; and similarly I will not give a woman a pessary to cause an abortion.
- In purity and according to divine law will I carry out my life and my art.
- I will not use the knife, even upon those suffering from stones, but I will leave this to those who are trained in this craft.
- Into whatever home I go, I will enter them for the benefit of the sick, avoiding any voluntary act of impropriety or corruption, including the seduction of women or men, whether they are free men or slaves.
- Whatever I see or hear in the lives of my patients, whether in connection with my professional practice or not, which ought not to be spoken of outside, I will keep secret, as considering all such things to be private.
- So long as I maintain this Oath faithfully and without corruption, may it be granted to me to partake of life fully and the practice of my art, gaining the respect of all men for all time.However, should I transgress this Oath and violate it, may the opposite be my fate". Translated by Michael North, National Library of Medicine, 2002. Updated 2012 (National Institutes of Health, and Human Services).

6.5.2 AMA Principles of Medical Ethics

The *American Medical Association* (**AMA**) has adopted a set of Ethical medical principles in 1957 and revised them in 1980 and 2001. These principles provide the standards of conduct which define the essentials of honorable behavior for the physician [9]. The AMA principles are the following:

1. "A physician shall be dedicated to providing competent medical care, with compassion and respect for human dignity and rights.
2. A physician shall uphold the standards of professionalism, be honest in all professional interactions, and strive to report physicians deficient in character or competence, or engaging in fraud or deception to appropriate entities.
3. A physician shall respect the law and also recognize a responsibility to seek changes in those requirements which are contrary to the best interest of the patient.
4. A physician shall respect the rights of patients, colleagues, and other health professionals, and shall safeguard patient confidences and privacy within the constraints of the law.
5. A physician shall continue to study, apply, and advance scientific knowledge, maintain a commitment to medical education, make relevant information available to patients, colleagues, and the public, obtain consultation and use the talents of other health professionals when indicated.
6. A physician shall, in the provision of appropriate patient care, except in emergencies, be free to choose whom to serve, with whom to associate and the environment in which to provide medical care.
7. A physician shall recognize a responsibility to participate in activities contributing to the improvement of the community and the betterment of public health.
8. A physician shall, while caring a patient, regard responsibility to the patient as paramount.
9. A physician shall support access to medical care for all people".

6.6 Concluding Remarks

In this chapter we provided fundamental elements of medical roboethics, particularly of robotic surgery ethics. The introduction of robots to healthcare functions adds complications to the assignation of liability. If something goes wrong there is the potential for damage to be caused to a patient, a medical practitioner or equipment. Ethical issues arise in connection to agency and responsibility, with a need to establish who is in control and at what point a duty arises. Robotic surgery ethics goes jointly with law and they are both considered jointly.

Each generation of surgeons inherits the medical ethical principles from the previous generation and has to obey dynamically changing legislation. Robotic surgery is now at a turning point having less to address a new technology and more to provide individual patient and global higher quality of life with coming new methods and practices of surgery. As it is evident, from the material of the chapter, the ethical dilemmas have dramatically increased in scope having to follow the coevolution of technology, society and ethics. Many surgeries such as orthopedic surgery, neurosurgery, pancreas surgery, cardiac surgery, microsurgery and general surgery benefit from the contribution of robotics. However, special training and experience together with high level assessment and surgery planning, and enhanced safety measures are required to provide normal consequentious care and state-of-art treatment.

An other area of medical care is *telemedicine* and *e-medicine* where, due to the use of the Internet and other communication/computer networks, there occur stronger ethical issues concerning data assurance and patient privacy [23, 24]. On the technical side, the problem of random time-delay over the internet should be faced, especially in telesurgery and e-telesurgery.

References

1. Pence GE (2000) Classic cases in medical ethics. McGraw-Hill, New York
2. Mappes TA, DeGrazia D (eds) (2006) Biomedical ethics. McGraw-Hill, New York
3. Szolovitz P, Patil R, Schwartz WB (1998) Artificial intelligence in medical diagnosis. Ann Int Med 108(1):80–87
4. Satava RM (2003) Biomedical, ethical and moral issues being forced by advanced medical technologies. Proc Am Philos Soc 147(3):246–258
5. Carrick P (2001) Medical ethics in the ancient world. Georgetown University Press, Washington, DC
6. Galvan JM (2003) On technoethics. IEEE Robot Autom Mag, Dec 2003
7. Medical Ethics, American Medical Association. www.ama-assn.org/ama/pub/category/2512.html
8. North M (2012) The Hippocratic Oath (trans). National Library of Medicine, Greek Medicine. www.nlm.nih.gov/hmd/greek/greek_oath.html
9. AMA's Code of Medical Ethics (1995) www.ama-assn.org/ama/pub/physician-resources/medical-ethics/code-medical-ethics.page?
10. Taylor BH et al (1996) Computer-integrated surgery. MIT Press, Cambridge
11. Gomez G (2007) Emerging technology in surgery: informatics, electronics, robotics. In: Towensend CM, Beauchamp RD, Evers BM (eds) Sabiston textbook of surgery. Saunders Elsevier, Philadelphia
12. Eichel L, Mc Dougall EM, Clayman RV (2007) Basics of laparoscopic urology surgery. In: Wein AJ (ed) Campbell-Walsh urology. Saunders Elsevier, Philadelphia
13. Jin L, Ibrahim A, Naeem A, Newman D, Markarov D, Pronovost P (2011) Robotic surgery claims on United States hospital websites **33**(6):48–52
14. Himpens J, Leman G, Cadiere GB (1998) Telesurgical laparoscopic cholecystectomy. Surg Endosc 8(12):1091
15. Marescau J, Leroy J, Gagner M et al (2001) Transantlantic robot-assisted tele-surgery. Nature 413:379–380
16. Satava RM (2002) Laparoscopic surgery, robots, and surgical simulation: moral and ethical issues. Semin Laparoscopic Surg 9(4):230–238

17. Mavroforou A, Michalodimitrakis E, Hatzitheo C, Giannoukas A (2010) Legal and ethical issues in robotic surgery. Int Angiol 29:75–79
18. Rogozea L, Leasu F, Rapanovici A, Barotz M (2010) Ethics, robotics and medicine development. In: Proceedings of 9th WSEAS international conference on signal processing, robotics and automation (ISPRA'10). University of Cambridge, England
19. Kemp DS (2012) Autonomous cars and surgical robots: a discussion of ethical and legal responsibility. Legal Analysis and Commentary from Justia, Verdict. http://verdict.justia.com/2012/11/19/autonomous-cars-and-surgical-robots
20. The European Group on Ethics makes public in Brussels its opinion on the "Ethical aspects of patenting inventions involving human stem cells", Brussels, 7 May 2002. http://europa.eu.int/comm/european_group_ethics
21. Sumulong C (2010) Are robotics a good idea for cardiac surgeons? Medical Robotics Magazine, 8 Mar 2010 (Interview with Henry Louie; MD, 18 Jan 2010. Interview with Dr. Mark Grattan, MD, 9 Nov 2009)
22. Robotic Surgery-Information Technology (2012) Further research for robotic surgery. http://kfrankl.blogspot.gr/2012/04/further-research-for-robotic-surgery.html
23. Merrel RC, Doarn CR (2003) Ethics in telemedicine research. Telemed e-health 15(2):123–124
24. Silverman R (2003) Current legal and ethical concerns in telemedicine and e-medicine. J Telemed Telecare **9**(Suppl. 1): 67–69

Chapter 7
Assistive Roboethics

Character is that which reveals moral purpose, exposing the class of things a man chooses and avoids.

Aristotle

The first step in the evolution of ethics is a sense of solidarity with other human beings.

Albert Schweitzer

7.1 Introduction

Assistive robots are called the robots that are designed for *people with special needs* **(PwSN)** in order to assist them to improve their mobility and attain their best physical and/or social functional level. Thanks to the progress in intensive care and early admissions in rehabilitation centers the number of severely injured people that survive is increasing, but very often with severe impairments. As a result, physical therapists have daily to care an increasing number of multi-handicapped persons with high-level dependence. A classification of PwSN was provided in Sect. 4.4.4 (PwSN with loss of upper limb control, PwSN with loss of lower limb control, PwSN with loss of spatio-temporal orientation). To design robots that can assist PwSN, e.g., the motor, disabled persons, roboticists should have a good knowledge of the context within which these robots are to function. Patients with loss of lower limbs are typically paraplegic persons (due to spinal cord injury, tumor, or degenerative disease). These people with full upper-limb power become able to move with classical manually propelled wheelchairs, walking aids, etc.

People with loss of upper-limb control can function with the help of manipulation robotic aids, and control a wheelchair using an appropriate joystick command. Typical cases are *tetraplegic* patients (due to cervical spinal cord injury) and *quadriplegic* due to other pathologies leading to motor deficit of the four limbs. Other people can be secondarily and/or provisionally unable to perform any effort toward autonomy (ageing, cardiac, arthropathy, spasticity, myopathy, poliomyelitis, etc.). Patients of this class can benefit by the use of semi-autonomous mobility aids via proper

S.G. Tzafestas, *Roboethics*, Intelligent Systems, Control and Automation: Science and Engineering 79, DOI 10.1007/978-3-319-21714-7_7

human-machine interface. People with loss of spatio-temporal orientation suffer from mental and neuropsychological impairments, vigilance disorders due traumatic brain injuries, stroke, ageing or visual deficiencies. These patients are unable to make any effort towards autonomy, and need intelligent robot-based mobility with high degree of robotic autonomy. Semi-autonomous navigation may be applicable subject to the capability of the system to switch to the autonomous mode if an inconsistent command is given or complete loss of orientation occurs [1]. A person's mobility skills should be carefully evaluated before selecting the type of wheelchair that should be used with optimal results. These include the patient's ability to functionally ambulate, propel a manual wheelchair and/or operate a powered wheelchair safely and efficiently through the environment. The medical status of the patient should be identified and taken into account. In general, the medical status of a person should include: (i) Primary diagnosis and prognosis, (ii) past medical history, (iii) past surgical history on all body parts that would affect mobility or seating, (iv) future medical/therapeutic intervention planned or considered, (v) rehabilitation measures needed to be taken, and (vi) the patient's medications and allergies.

The purpose of this chapter is to provide basic information on assistive robotics and ethics. Specifically, the chapter:

- Discusses a set of assistive robotic devices for people with impaired upper limbs/hands and lower-limbs (wheelchairs, orthotic devices, prosthetic devices, rehabilitation robots). This material is intended to help the reader to appreciate better the ethical issues involved.
- Outlines the fundamental ethical principles and guidelines of assistive robotics including the ethical codes of **RESNA** (The Rehabilitation Engineering and Assistive Technology Society), and **CRCC** (The Canadian Commission on Rehabilitation Counselor Certification).

7.2 Assistive Robotic Devices

Assistive robotic devices include:

- Assistive robots for people with impaired upper limbs and hands.
- Assistive robots for people with impaired lower limbs.
- Rehabilitation robots (upper limb or lower limb).
- Orthotic devices.
- Prosthetic devices.

7.2.1 Upper Limb Assistive Robotic Devices

Robotic devices in this class are designed to assist persons with severe disabilities to perform everyday functions such as eating, drinking, washing, shaving, etc. Two

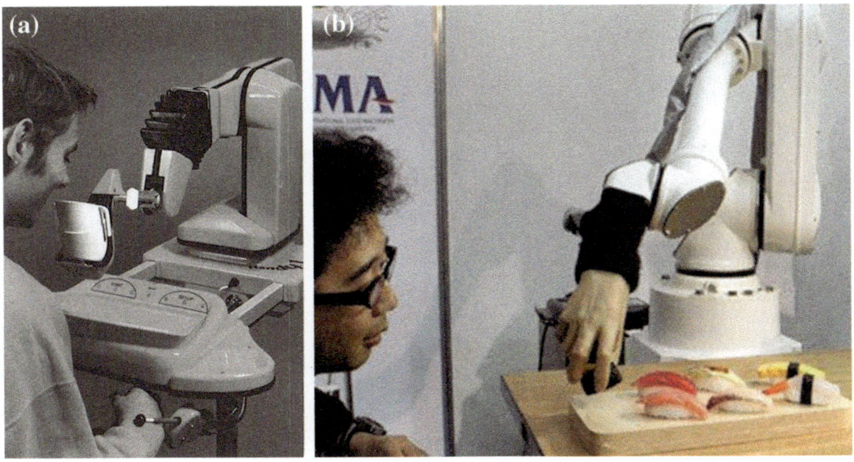

Fig. 7.1 **a** The Handy 1 multifunctional robot, **b** a modern service-assistive robot with 5-finger human-like hand (Koii Sasahara/AP). *Source* (**a**) http://www.emeraldinsight.com/content_images/fig/0490280505001.png, (**b**) http://www.blogcdn.com/www.slashfood.com/media/2009/06/sushi-hand-food-425rb061009.jpg

such robots are the unifunctional *'My Spoon' robot* developed to help those who need assistance to eat [2], and *Handy 1* which is a multifunctional robot that assists in most upper arm everyday functions (Rehab Robotics Ltd, UK) [3]. The Handy 1 robot was developed at the University of Keele in 1987 and can interact with the user, also having preprogrammed motions that help in fast completion of tasks. Handy 1 is one of the first assistive robots of this kind. Presently there are more intelligent robots and mobile manipulators that assist or serve people with motion disabilities. Figure 7.1 shows the *Handy 1* robot and a more advanced robot that can assist people with upper-hand disabilities.

Another robot designed to help people with upper-limb dysfunctions is the MANUS manipulator which is a robot directly controlled by the individual requiring each movement of the robotic manipulator to be controlled by a corresponding movement action of the person [4]. It has been successfully applied by people with muscular dystrophy and similar conditions that lead to muscle weakness. This principle can also allow the user to have a physical sense of the environment feedback as forces to the point of interaction. The most recent version of MANUS is a 6 + 2 degrees of freedom manipulator controlled by joysticks, chin controllers, etc. The MANUS robot can be mounted on a wheelchair as shown in Fig. 4.26a for the FRIEND powered wheelchair [5].

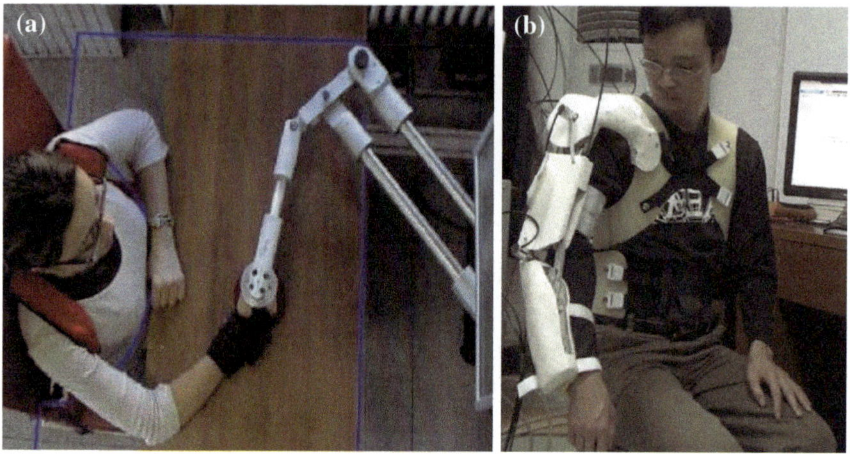

Fig. 7.2 **a** A robotic upper-limb rehabilitation robotic arm, **b** an exoskeleton arm rehabilitation robotic device that helps stroke victims. *Source* (**a**) http://www.robotliving.com/wp-content/uploads/20890_web.jpg, (**b**) http://www.emeraldinsight.com/content_images/fig/0490360301011.png

7.2.2 *Upper Limb Rehabilitation Robotic Devices*

These devices are used for the evaluation and therapy of arms impaired as consequence of stroke [1, 6]. These devices promise good assistance for improvement of motor impairments. However, to have the best results a deep evaluation is required for each particular case in order to use the most appropriate device. Two devices for this purpose are: ARM Guide, and Bi-Manu-Truck. The ARM Guide has a single actuator, and the motion of the patient's arm is constrained to a linear path that can be oriented within the horizontal and vertical planes. This device was verified to offer quantifiable benefits in the neuro-rehabilitation of stroke persons. In general, using therapeutic robots in the rehabilitation process, specific, interactive and intensive training can be given. Figure 7.2a shows a robotic arm helping stroke victims, and Fig. 7.2b shows an exoskeleton orthotic robot for upper-limb rehabilitation.

7.2.3 *Lower Limb Assistive Robotic Mobility Devices*

These devices include robotic wheelchairs and walkers. Two robotic wheelchairs are depicted in Fig. 4.17a, b. The "FRIEND" wheelchair of Fig. 4.17a was developed at the University of Bremen (IAT: Institute of Automation) and offers increased control functionality to the disabled user [5]. It consists of an electric wheelchair equipped with a robotic manipulator which has at its endpoint a fingered

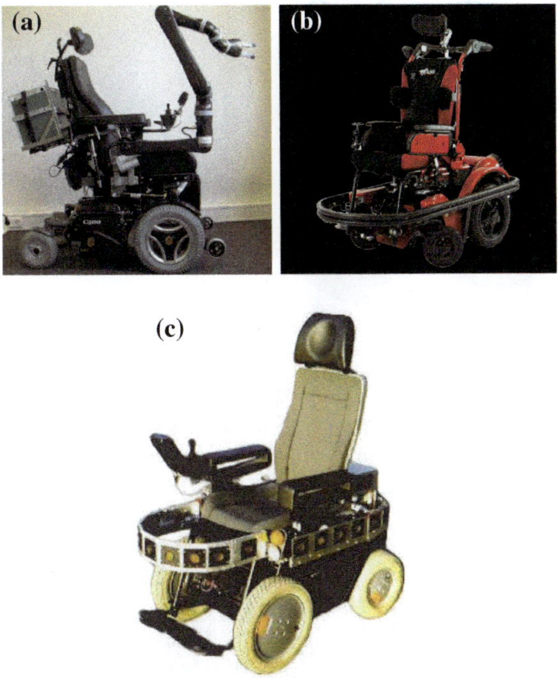

Fig. 7.3 **a** Wheelchair with mounted manipulator, **b** the SMART children-wheelchair, **c** the "Rolland" autonomous wheelchair. *Source* (**a**) http://www.rolstoel.org/prod/images/kinova_jaco_2.jpg, (**b**) http://www.smilerehab.com/images/Smart-chair-side.png, (**c**) http://www.informatik.uni-bremen.de/rolland/rolland.jpg

human-like hand. Both devices are controlled by a computer via a suitable human-machine interface. Three other wheelchairs are shown in Fig. 7.3. The first is a wheelchair similar to "FRIEND" wheelchair that can assist the user in simple every-day functions. The second is the "SMART" wheelchair which is suitable for children with severe multiple disabilities. It is available commercially (Call Centre and Smile Rehab. Ltd.) [7]. If collision occurs, the wheelchair stops and initiates an avoiding action (i.e., stop, back-off, turn around the obstacle, etc.) as it may be required. This wheelchair can be driven by single or multiple switches, scanning direction detector, and a joystick. The third is the "ROLLAND" intelligent wheelchair of the University of Bremen which can work autonomously with the help of a ring of ultrasonic sensors (range detectors), an adaptive sonar firing strategy/algorithm, an obstacle avoidance subsystem, a "stopping-in-time" safety scheme, and proper "mobility assistants" (driving assistant, route assistant).

Some other intelligent (autonomous/semiautonomous) wheelchairs are:

- *Navchair* (University of Michigan, U.S.A) [8]
- *VAHM* (University of Metz, France) [9]
- *MAID* (FAW Ulm, Germany) [10]

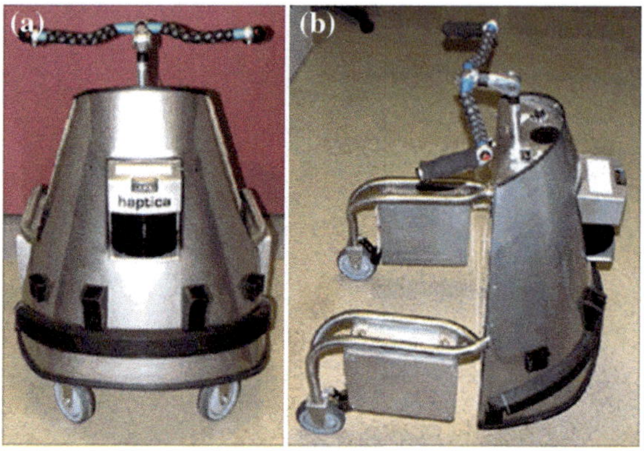

Fig. 7.4 The VA-PAMAID walker **a** front view, **b** side view. The walker has three control modes: manual, automatic, and park. *Source* http://www.rehab.research.va.gov/jour/03/40/5/images/rentf01.gif

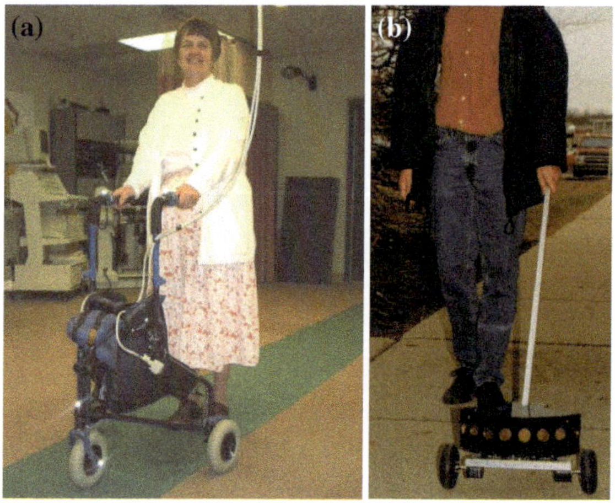

Fig. 7.5 **a** Assistive robotic walker of the University of Virginia Medical Center **b** guide cane of the University of Michigan. *Source* (**a**) hhttp://www.cs.virginia.edu/~gsw2c/walker/walker_and_user.jpg, (**b**) http://www-personal.umich.edu/~johannb/Papers/Paper65/GuideCane1.jpg

- *ASIMOV* (Lund University, Sweden) [11]
- *INRO* (FHWeingarten, Germany) [12]

Robotic Walkers Robotic walkers are designed for people who have some basic physical and mental abilities to perform a task, but they can perform it inefficiently and unsafely. A walker can help a person to navigate and avoid collisions with obstacles, thus helping to reduce health costs and increase the quality of care and independency of handicapped people. In contrast to wheel chairs (hand driven or powered), walkers seek to help people who can and want to walk.

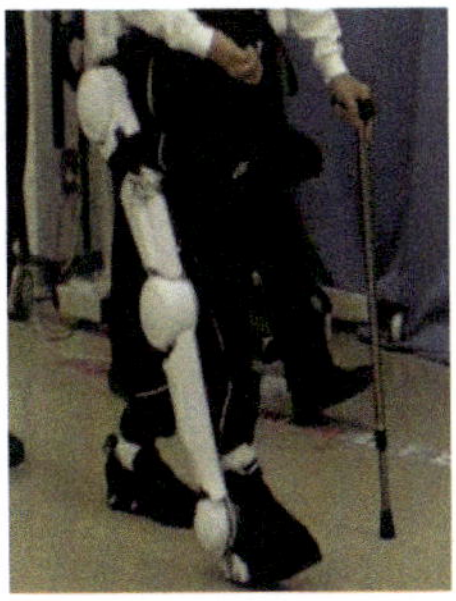

Fig. 7.6 An exoskeleton robotic walking device. *Source* http://www.hindawi.com/journals/jr/2011/759764.fig.009.jpg

A well known robotic walker is the "Veteran Affairs Personal Adaptive Mobility Aid" **(VA-PAMAID)**. The original prototype walker was developed by Gerald Lacey while at Trinity College in Dublin. The commercialization of VA-PAMAID is done jointly with *Haptica Company* [13]. The walker is shown in Fig. 7.4. Another robotic walker was developed at the Virginia University Medical Center (Fig. 7.5a). Figure 7.5b shows the University of Michigan Guide Cane.

7.2.4 *Orthotic and Prosthetic Devices*

Orthotic devices are used to assist or support a weak and ineffective muscle or limb. Typical orthotic devices take the form of an exoskeleton, i.e., a powered anthropomorphic suit that is worn by the patient. One of the early orthotic devices is the "*wrist-hand orthotic*" **(WHO)** device which uses shape memory alloy actuators for providing a grasping function for quadriplegic persons [14]. Exosceleton devices

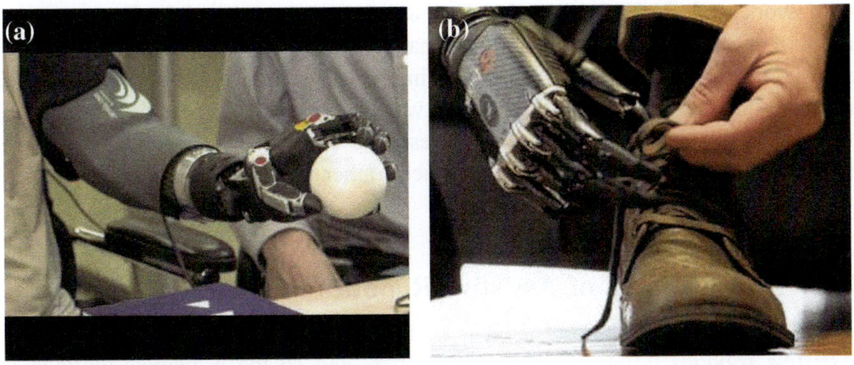

Fig. 7.7 Two prosthetic upper-limb/hand robotic devices. *Source* (**a**) http://i.ytimg.com/vi/VGcDuWTWQH8/0.jpg, (**b**) http://lh5.ggpht.com/-Z7z0l844hhY/UVkhVx5Uq8I/AAAAAAAAB08/AajQjbtsK8o/The%252520BeBionic3%252520Prosthetic%252520Hand%252520Can%252520Now%252520Tie%252520Shoelaces%25252C%252520Peel%252520Vegetables%252520and%252520Even%252520Touch%252520Type.jpg

Fig. 7.8 Prosthetic lower-limb device. *Source* http://threatqualitypress.files.wordpress.com/2008/11/prosthetics-legs.jpg

have links and joints, corresponding to those of the human, and actuators. An arm exoskeleton is shown in Fig. 7.2b. A leg exoskeleton has the form of Fig. 7.6.

Prosthetics are devices that are used as substitutes for missing parts of the human body. These devices are typically used to provide mobility or manipulation when a limb is lost (hence the name *artificial limbs*). A representative prosthetic device is the *Utach Arm* (Motion Control Inc., U.S.A.) [15]. It is a computer controlled, above the elbow, prosthesis which uses feedback from *electromyography* **(EMG)** sensors that measure the response of the muscle to nervous stimulation electrical activity within muscle fibers. Other prosthetic systems can determine the intended action on the human so that the prosthetic device can be properly controlled. Figure 7.7 shows two prosthetic upper-limb devices, and Fig. 7.8 shows an exoskeleton walking device.

Artificial limbs are designed by professionals who specialize in making prosthetic limbs. Most people who wear prosthetic limbs are able to return to their previous activity levels and lifestyles. However, this can only be accomplished with hard work and determination. In general, acceptance or denial of an assistive robotic device depends on what the machine can do or cant' do. Moreover, social/ethical factors may also determine acceptability. Several studies have revealed that there is a clear difference between elderly and younger or handicapped in accepting or rejecting assistive devices. The elderly tend in general to refuse technical innovation, and prefer human help rather than technological help.

7.3 Ethical Issues of Assistive Robotics

Assistive robotics is part of medical robotics, and so the principles outlined in Chap. 6 are applicable. Any assistive robot or device has the potential to be beneficially used or to be misused. In any case an assistive device may be beneficial in some respects, but may also have costs to the PwSNs or their caregivers. The basic ethical questions surrounding the development of assistive robotics are focused on human dignity, human relations, protection from physical/bodily harm, and the management of the health evaluation and other personal data.

The six part medical ethics guide (*Georgetown mantra*) are also valid here, namely (see Sect. 6.2):

- Autonomy
- Non-maleficence
- Beneficence
- Justice
- Truthfulness
- Dignity

Other aspects that belong to the responsibility of the doctor/caregiver include:

- Confidentiality/Privacy
- Data integrity
- Clinical accuracy
- Quality
- Reliability

All assistive technology programs should incorporate ethics statements into administrative policies and comply with professional codes of assistive ethics. Unfortunately it was proved in practice that in many cases the moral implications were either ignored or made subservient to a more pressing commercial need.

The first thing required for the decision of a particular assistive program is an accurate clinical/physical/psychological evaluation of the person to be rehabilitated. Some aspects that should be carefully considered are:

- Selection of the most proper device (economically allowed and affordable).
- Assurance that the selected device is not used for doing things that a person is still able to do for him/herself (which will probably make the problems worse).
- Use of technological solutions that may not restrict freedom or privacy with full involvement and consent of the person.
- Consideration of assistive technology that could be used to help a person for doing things he/she finds harder to do.
- Assurance of safety which is of great importance.

The *Rehabilitation Engineering and Assistive Technology Society:* **RESNA** (U.S.A. 2012) has released the following ethical code for the use of assistive technology by its members [16]:

- Hold paramount the welfare of persons served professionally.
- Practice only in their area(s) of competence and maintain high standards.
- Maintain the confidentiality of privileged information.
- Engage in no conduct that constitutes a conflict of interest or that adversely reflects on the association and, more broadly, on professional practice.
- Seek deserved and reasonable remuneration for services.

- Inform and educate the public on rehabilitation/assistive technology and its applications.
- Issue public statements in an objective and truthful manner.
- Comply with the laws and policies that guide professional practice.

The *Canadian Commission on Rehabilitation Counselor Certification*: **CRCE** (2002) has developed the following code which is applicable to any professional rehabilitation therapist [17]:

- Advocate for people with disabilities by addressing physical and attitudinal barriers to full participation in society.
- Empower consumers by providing them with the information they need to make informed choices about the use of the services.
- Adhere to the principles of confidentiality and privacy.
- Have professional responsibility.
- Maintain professional relationships.
- Provide an excellent evaluation, assessment, and interpretation.
- Make use of teaching, supervision, and training.
- Perform research and publish.
- Provide technology and distance counseling.
- Have correct business practices.
- Resolve ethical issues.

A general four-level ethical decision-making model for assistive/rehabilitation robotics and other assistive technologies is the following [18]:

Level 1: Selection of Proper Assistive Devices

This is the 'client professional relationship' level. Consumers should be provided the appropriate rehabilitation services. Employing inappropriate and counter-productive devices and services is a violation of the non-maleficence ethical rule. The rules of beneficence, justice and autonomy should also be respected at this level.

Level 2: Competence of Therapists

This is the 'clinical multidisciplinary' level, and is implemented via the effective relationship between practitioners. Some therapists may be more competent in using rehabilitation devices than others. Here again, the principles of non-maleficence, truthfulness, beneficence, justice, and autonomy should be adhered.

Level 3: Efficiency and Effectiveness of Assistive Devices

This is the 'institutional/agency' level. Institutions and agencies are legally and ethically obliged to guarantee the provision of assistive devices/services efficiently and effectively. Efficiency means the use of cost effective devices that are reliable and efficient. Here, the justice ethical rule is of highest priority, i.e., the PwSNs must have their rehabilitation service needs met. Professionals of institutions/agents should be well educated in the up-to-date assistive technologies.

Level 4: Societal Resources and Legislation

This is the 'societal and public policy' level which is institutionalized by the legislative constituent relationship. Best practices rehabilitation interventions should be employed for all aspects. User, agency, and societal resources should be properly exploited in order to access the best available technologies. The driving force should always be the quality of life improvement through the use of resources with maximum efficiency in conformity with the justice and autonomy principles.

Other ethical considerations in assistive robotics and technology include:

- The establishment of a support professional/patient relationship.
- Standardization of practice and patient privacy.
- Reimbursement for assistive technology exchanges.
- Avoidance of social and health malpractice (with licensure and disciplinary concerns).
- An effort should be made by manufacturers to produce affordable devices for the patients and insurance bodies, based on market research.
- Consideration of the views of other professional care-givers about the proposed/selected rehabilitation action and the consequences of doing or not doing it.

7.4 Concluding Remarks

Robotic and other assistive devices for the impaired can provide valuable assistance to their users. The connection between the device and the user/PwSN is the key to this. This depends largely on the evaluation that includes both clinical needs and technological availability. This suggests that medical and technological assistive evaluations should be continuously refined to meet ethical/social standards, and to assure full acceptance by the users who should be truly convinced that the proposed assistive device(s) (active or passive) will help them to increase motion autonomy and improve their quality of life. In this chapter we have outlined major representative robotic assistive devices for upper and lower limb functioning and therapy/rehabilitation. The available literature on assistive technologies and assistive robotics, in particular, is vast. The same is true for the ethical/moral issues. The chapter has provided a discussion of the basic ethical principles and guidelines which assure a successful and ethical employment and exploitation of assistive robots, compatible with the general ethical principles of medical practice. For further information on assistive roboethics the reader is referred to [19–30].

References

1. Katevas N (ed) (2001) Mobile robotics in healthcare (Chapter 1). IOS Press, Amsterdam
2. Soyama R, Ishii S, Fukuse A (2003) The development of meal-assistance robot: My spoon. In: Proceedings of 8th international conference on rehabilitation robotics (ICORR'2003), pp 88–91. http://www.secom.co.jp/english/myspoon
3. Topping M (2002) An overview of the development of Handy 1: a rehabilitation robot to assist the severely disabled. J Intell Rob Syst 34(3):253–263. www.rehabrobotics.com
4. MANUS Robot. www.exactdynamics.nl
5. http://ots.fh-brandenburg.de/downloads/scripte/ais/IFA-Serviceroboter-DB.pdf
6. Speich JE, Rosen J (2004) Medical robotics. Encycl Biomaterials Biomech Eng. doi:10.1081/E-EBBE-120024154
7. http://callcentre.education.ed.ac.uk/downloads/smartchair/smartsmileleaflet. Also http://www.smilerehab.com/smartwheelchair.html
8. Levine S, Koren S, Borenstein J (1990) NavChair control system for automatic assistive wheelchair navigation. In: Proceedings of the 13th annual RESNA conference, Washington
9. Bourhis G, Horn O, Habert O, Pruski A (2001) An autonomous vehicle for people with motor disabilities (VAHM). IEEE Robot Autom Mag 8(1): 20–28
10. Prassler E, Scholz J, Fiorini P (2001) A robotic wheelchair for crowded public environments (MAid). IEEE Robot Autom Mag 8(1):38–45
11. Driessen B, Bolmsjo G, Dario P (2001) Case studies on mobile manipulators. In: Katevas N (ed) Mobile robotics in healthcare (Chapter 12). IOS Press, Amsterdam
12. Shilling K (1998) Sensors to improve the safety for wheelchair users, improving the quality of life for the European citizen. IOS Press, Amsterdam, pp 331–335
13. Rentsehler AJ, Cooper RA, Blasch B, Boninger BL (2003) Intelligent walkers for the elderly: performance and safety testing of VA-PAMAID robotic walker. J Rehabil Res Dev 40(5):423–432
14. Makaran J, Dittmer D, Buchal R, MacArthur D (1993) The SMART(R) wrist-hand orthosis (WHO) for quadriplegics patients. J Prosthet 5(3):73–76
15. http://www.utaharm.com
16. RESNA code of ethics. http://resna.org/certification/RESNA_Code_of_Ethics.pdf
17. www.crccertification.com/pages/crc_ccrc_code_of_ethics/10.php
18. Tarvydas V, Cottone R (1991) Ethical response to legislative, organizational and economic dynamics: a four-level model of ethical practice. J Appl Rehabil Couns 22(4):11–18
19. Salvini P, Laschi C, Dario P (2005) Roboethics in biorobotics: discussion and case studies. In: Proceedings of IEEE international conference on robotics and automation: workshop on roboethics, Rome, April 2005
20. Garey A, DelSordo V, Godman A (2004) Assistive technology for all: access to alternative financing for minority populations. J Disabil Policy Stud 14:194–203
21. Cook A, Dolgar J (2008) Cook and Hussey's assistive technologies principles and practices. Mosby Elsevier, St. Louis
22. RESNA. Policy, legislation, and regulation. www.resna.org/resources/policy%2C-legislation%2C-and-regulation.dot
23. WHO. world health organization, world report on disability, 2011. www.who.int/disabilities/world_report/2011/report/en/index.html
24. Zwijsen SA, Niemejer AR, Hertogh CM (2011) Ethics of using assistive technology in the care for community-dwelling elderly people: an overview of the literature. Aging Ment Health 15(4):419–427
25. T.R.A.o.E.A. Systems, social, legal and ethical issues. http://www.raeng.org.uk/policy/engineering-ethics/ethics
26. RAE (2009) Autonomous systems: social, legal and ethical issues. The Royal Academy of Engineering. www.raeng.org.uk/societygov/engineeringethics/events.html

27. Peterson D, Murray G (2006) Ethics and assistive technology service provision. Disabil Rehabil Assistive Technol 1(1–2):59–67
28. Peterson D, Hautamaki J, Walton J (2007) Ethics and technology. In: Cottons R, Tarvydas V (eds) Counseling ethics and decision making. Merill/Prentice Hall, New York
29. Zollo L, Wada K, Van der Loos HFM (Guest eds) (2013) Special issue on assistive robotics. IEEE Robot Autom Mag 20(1):16–19
30. Johansson L (2013) Robots and the ethics of care. Int J Technoethics 4(1):67–82. www.irma-international.org/article/robots-ethics-care/77368

Chapter 8
Socialized Roboethics

The top two awards don't even go to robots.
Chuck Gozdzinski

Live one day at a time emphasizing ethics rather than rules.
Wayne Dyer

8.1 Introduction

Service robots (or "*serve us robots*") are robots that function semi-autonomously or autonomously to carry-out services (other than the jobs performed in a manufacturing shop floor) that contribute to the well-being of humans. These robots are capable of making partial decisions and work in real dynamic or unpredictable environments to accomplish desired tasks.

Joseph Engelberger, the father of modern robotics, predicted that some day in the not too distant future, service robots will be the widest class of robots outnumbering the industrial robots by several times. A working definition of *service robot* was given by **ISRA** (**I**nternational **S**ervice **R**obot **A**ssociation) which states: "Service robots are machines that sense, think, and act to benefit or extend human capabilities, and to increase human productivity". Of course, the philosophical and practical meaning of the words *robot*, *machine*, and *think* need to be investigated and properly interpreted as explained in many places of this book. Clearly, if the public at large is to be the final end-user of service robots, the issue of what roboticists can do to inform, educate, prepare and involve the societal moral norms, need to be seriously considered.

A phenomenon that must be properly faced is the irreversible aging of population worldwide. Several statistical studies of international organizations conclude that the number of young adults for every older adult is decreasing dramatically. It is anticipated that the percentages of people over 85, in the next decade(s) should exhibit a big increase of elderly people and a shortage of personnel to take care of them. Therefore it seems a necessity to strengthen the efforts for developing

S.G. Tzafestas, *Roboethics*, Intelligent Systems, Control and Automation: Science and Engineering 79, DOI 10.1007/978-3-319-21714-7_8

assistive and socialized service robots, among others, that could provide continuous care and entertainment of the elderly for improving their quality of life at the final period of their life.

This chapter is devoted to the study of socialized (entertainment, companion, and therapeutic) robots. Specifically, the chapter:

- Provides a typical classification of service robots.
- Presents the various definitions of socialized robots discussing briefly the desired features that should have.
- Provides a number of representative examples of socialized (anthropomorphic, pet-like) robots.
- Discusses the fundamental ethical issues of socially assistive robots.
- Reviews three case studies concerning children-robot and elderly-robot interactions for autistic children and dementia elders.

8.2 Classification of Service Robots

Service robots possess several levels of artificial intelligence and intelligent human-machine interfaces and interaction. Therefore, as argued in [1], robots are classified as follows:

- Robots serving as tools
- Robots serving as cyborg extensions
- Robots as avatars
- Robots as sociable partners

Robots serving as Tools The humans regard the robots as tools that are used to perform desired tasks. Here, industrial robots, teleoperated robots, household robots, assistive robots and in general all robots that need to be supervised, are included.

Robots serving as cyborg extensions The robot is physically connected with the human such as the person accepts it as an integrated part of his/her body (e.g., the removal of a robotic leg would be regarded by the person as a partial and temporary amputation).

Robots as avatars The individual projects his/herself via the robot for communicating with another far away individual (i.e., the robot gives a sense of physical and social presence of the humans interacting with it).

Robots as social partners The interaction of a human with a robot appears to be like interacting with another socially responsive creature that cooperates with him/her as a partner (it is remarked that at present full human-like social interaction has not yet been achieved).

In all cases there is some degree of shared control. For example, an autonomous vehicle can self-navigate. A cyborg extension (prosthetic device) might have basic reflexes based on proper feedback (e.g., temperature feedback to avoid damage of the cyborg, or touch/tactile feedback to grasp a fragile object without breaking it).

A robot avatar is designed to coordinate speech, gesture, gaze, and facial expression, and shows them at the right time to the proper person. Finally, a robot partner mutually controls the dialog and the exchange of talking with the cooperating human.

On the basis of the above we can say that all robots are service robots. In Chaps. 6 and 7 we have discussed the ethical issues of medical robots (surgical robots and assistive robots). Here, we will mainly be concerned with the so called *socialized robots* which are used for therapy, entertainment or as companions. According to the *American Food and Drug Administration*, socialized robots are labeled as *Class 2 medical devices*, like powered wheelchairs (e.g., stress relievers that calm elderly dementia patients).

(a)

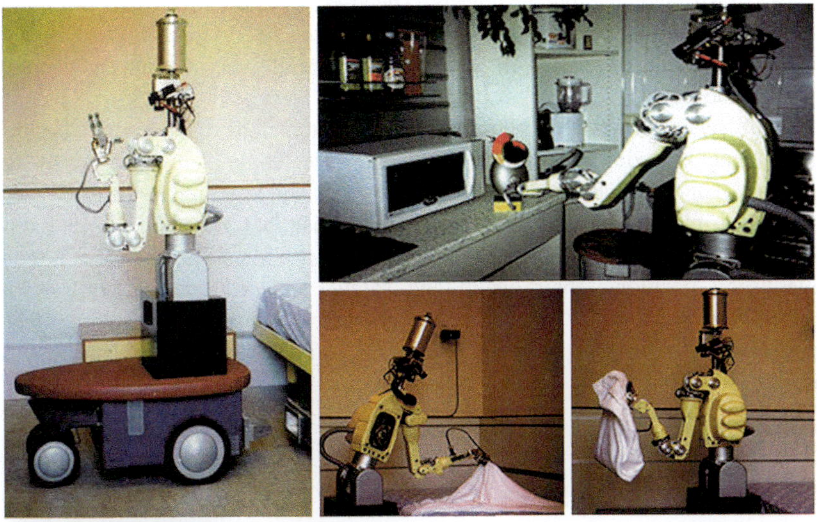

(b)

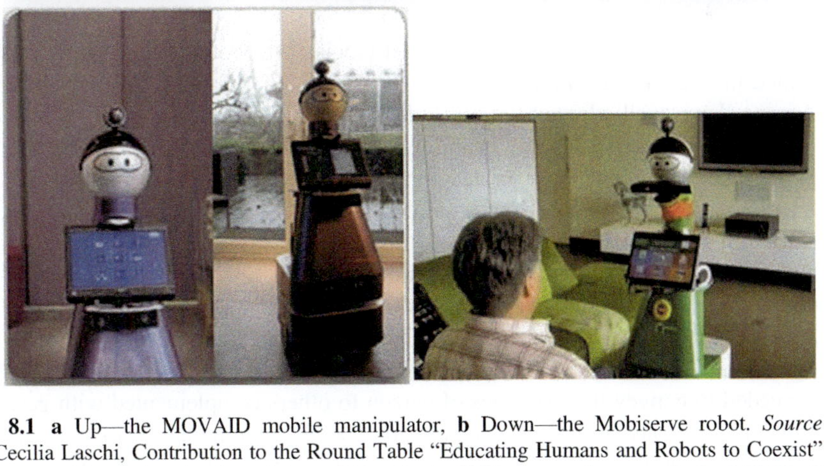

Fig. 8.1 **a** Up—the MOVAID mobile manipulator, **b** Down—the Mobiserve robot. *Source* (**a**) Cecilia Laschi, Contribution to the Round Table "Educating Humans and Robots to Coexist" Italy–Japan Symposium 'Robots Among Us', March, 2007, www.robocasa.net, (**b**, *left*) http://www.computescotland.com/images/qZkNr8Fo7kG90UC4xJU4070068.jpg; (**b**, *right*) http://www.vision-systems.com/content/vsd/en/articles/2013/08/personalized-vision-enabled-robots-for-older-adults/_jcr_content/leftcolumn/article/headerimage.img.jpg/1377023926744.jpg

Other "class 2 medical robots" are assistive robots (typically mobile robots and mobile manipulators) that help impaired people to everyday functions, such as eating, drinking, washing, etc. or for hospital operations. Examples of such robots are: the "*Care-0-Bot 3*" robot discussed in Sect. 4.4.3, the robots of Fig. 4.20, the *My Spoon* robot, the *Handy 1* robot (Fig. 7.1a), and the *MOVAID* and *Mobiserve* robots (Fig. 8.1).

MOVAID This robot (**M***obility and* **A***ctivity Assistance System for the Disabled*) is a mobile manipulator developed at the Sant'Anna Higher School (Italy). The design philosophy behind MOVAID was "*design for all*" and "*user oriented*". The system is accompanied by several personal computers located at the places of activities (kitchen, bedroom, TV room, etc.) and is able to navigate, avoid obstacles, dock, grasp, and manipulate objects. The user gives commands to the robot via "graphical user interfaces" (**GUIS**) running on the fixed workstation. Visual feedback from on-board cameras is given to the user allowing monitoring of what the robot is doing.

Mobiserve This is a mobile wheeled semi-humanoid robot equipped with sensors, cameras, audio and a touch screen interface. Some of its tasks are to remind persons that they have to take their medicine, suggest they have to drink their favorite drink or propose them to go a walk or visit their friends if they are at home longer than some time. Other tasks include smart home operations for monitoring the users' positions, their health and safety, and alert emergency services if something goes wrong. Furthermore by wearing smart clothes the robot can monitor vital signs such as sleeping patterns and detection if the wearer has fallen down, as well as eating and drinking patterns.

8.3 Socialized Robots

A partial list of properties that socialized (sociable) robots must possess was presented in Sect. 4.8. In the literature, several types of sociable robots are available. Among them, well adopted types are the following [2–5].

Socially evocative Robots of this type rely on the human tendency to anthropomorphize and capitalize on feelings evoked, when humans nurture, care or interact with their creation. Very common socially evocative robots are toy-like or pet-like entertainment robots (such as the robots shown in Fig. 4.42).

Socially commutative Robots of this type use human like social cues and communication patterns that make the interactions more natural and familiar. Socially commutative robots are capable of distinguishing between other social agents and several objects in their environment. Here, a sufficient social intelligence is needed to convey the messages of person to others complemented with gestures, facial expressions, gaze, etc. A class of socially commutative robots includes museum tour guides which convey the interaction information using speech and/or reflexive facial expressions.

Socially responsive These robots can learn through human demonstration with the aid of a training cognitive model. Their goal is to satisfy internal social aims (drives, emotions, etc.). They tend to be more perceptive of human social patterns, but they are passive, i.e., they respond to individual's attempts to interact with them without been able to proactively engage with people in satisfying internal social aims.

Sociable These robots proactively engage with people in order to satisfy internal social aims (emotions, drives, etc.) including both the person's benefit and its own benefit (e.g., improve its performance).

Socially intelligent These robots possess several capabilities of human-like social intelligence which are achieved using deep models of human cognition and social performance.

An alternative term for "*socially intelligent robot*" is the *term* "*socially interactive robot*" [6] where social interaction plays a dominant role in peer-to-peer *human-robot interfaces* (**HRI**s), different than standard HRIs used in other robots (e.g., teleoperation robots).

A list of capabilities possessed by "socially interactive robots" is the following [5, 6]:

- Express and/or perceive emotions.
- Communicate with high-level dialogue.
- Recognize other agents and learn their models.
- Establish and/or sustain social connections.
- Use natural patterns (gestures, gaze, and so on).
- Present distinctive personality and character.
- Develop and/or learn social competencies.

Sociable robots are also the robots that imitate humans. The two fundamental questions that have to be addressed for the design of such robots are [7]:

- How does a robot know what to imitate?
- How does a robot know how to imitate?

With reference to the first question the robot needs to detect the human demonstrator, observe his/her actions, and identify the ones that are relevant to the desired task, from those that are involved in the instructional/training process. This requires the robot's capability to perceive the human movement, and determine what is important to direct attention. The movement perception can be performed by 3-D vision or by using motion capture techniques (e.g., external worn exosceletons). The robot's attention is achieved by using attention models that selectively orient computational resources to areas that contain task related information. Human cues that have to be identified (by the vision system) are pointing, head pose, and gaze direction.

With reference to the second question the robot needs, after the action perception, to convert this perception into a sequence of its own motor motions such that to get the same result. This is called the "*correspondence problem*" [8]. In simple cases the correspondence problem can be solved a priori by using the learning-by-imitation paradigm. In complex cases the problem can be solved in the actuator space (representing perceived movements in joint based human arm movements) and transformed to the robot (e.g., as it is done in the *Sarcos Sen Suit* robot) [7]. Alternatively, the correspondence problem can be solved by representing the robot movements in task space terms and compare them with the observed human motion trajectory.

The robot's *social learning* involves several levels. Let **A** and **B** be two individuals (or groups). Then the social *bottom-up* learning levels are [9]:

- **Imitation**: **A** learns a new behavior of **B** and can perform it in the absence of **B**.
- **Goal emulation**: **A** can perform the observed end result of **B** with a different behavior.
- **Stimulus enhancement**: The attention of **A** is drawn to an object or location as a result of **B**'s behavior.
- **Exposure**: Both **A** and **B** face the same situation (due to that **A** is associated to **B**) and acquire similar behaviors.
- **Social facilitation**: **A** releases an intrinsic behavior as a result of **B**'s behavior.

It is emphasized that the above learning levels refer to dynamic and interactive social learning. Traditional *animatronic* devices (e.g., those used in entertainment parks) continuously replay movements that have been recorded (manually or by automatic devices) but they are not interactive. They cannot respond to environment's changes and cannot adapt to new situations.

8.4 Examples of Socialized Robots

Over the years many socialized robots have been developed in Universities, research institutes, and commercial robotic companies and manufacturers. These robots were designed so as to have a variety of capabilities that depend on their ultimate goal. Here, a few representative socialized robots will be briefly described, in order to give the reader a better feeling of what they do and how they do entertainment and therapeutic tasks.

8.4.1 Kismet

This *anthropomorphic robot head*, created at MIT, is shown in Fig. 4.32. Actually, Kismet is not intended to perform particular tasks. Instead, it was designed to be a robotic creature that has the ability to interact with humans physically, effectively and emotionally in order to finally learn from them. Kismet can elicit interactions with humans that need high-level rich learning features. This ability allows it to perform playful infant-like interactions that help infants/children to achieve and develop social behavior.

Particular capabilities of Kismet include:

- Recognition of expressive feedback like prohibition or praise.
- Regulation of interaction in order to create a proper environment.
- Taking turns in order to structure the learning facet.

Full technical and operational details of Kismet are provided in the book of its creator *Cynthia Breazeal* [3].

8.4.2 *Paro*

This is a robot developed by the Japan National Institute of Advanced Industrial Science and Technology **(AIST)** [10], and manufactured by the company "Intelligent Systems Co". Paro seems to be responding to its environment and humans interacting with it (e.g., petting, talking to it, etc.). Paro was designed such as its interaction with humans are aimed to mimic not actual interactions with a seal, but rather what a fantasy of interacting with a baby seal would be like. Paro's capabilities are acquired by a sophisticated internal system that incorporates microprocessors, tactile sensors, light-sensors, touch-sensitive whiskers, sound and voice recognition, and blinking eyes placed in a head that turns and seems to be tracking human motion and pay attention to a person interacting with it. Two snapshots of Paro are shown in Fig. 8.2.

The use of the seal which is an unfamiliar animal has the benefit that users have the impression that the robot acts exactly as the real animal. Interaction with familiar animals (dogs, cats) may be easily detected that differs from those of the real living animal. Paro reduces patients'/care givers' stress, improves the socialization of patients and caregivers, and improves relaxation and motivation. It responds positively to gentle touches and nice words that patients are drawn to it.

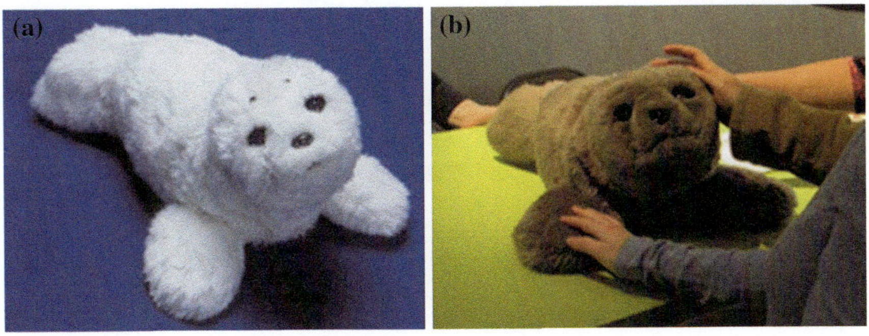

Fig. 8.2 The robot baby seal Paro. *Source* (**a**) http://gadgets.boingboing.net/gimages/sealpup.jpg, (**b**) http://www.societyofrobots.com/images/misc_robotopia_paro.jpg

8.4.3 CosmoBot

This robot has been developed to help children with disabilities [11]. Children interact with it for therapy, education and play. They control CosmoBot's movements and audio output via a set of gesture sensors and speech recognition. Several studies have verified that children with *cerebral palsy* who interacted with the robot improved their fitness level, e.g., they increased the strength of their quadriceps to the point at which they are within the statistical norm. Moreover, children with cerebral palsy achieved visible increase of strength and independent leg and torso function as a result of using a custom walker. CosmoBot is shown in Fig. 8.3. The robot has software that includes data tracking capabilities for automatic data recording from therapy sessions. The children are able to control the robot's head, arm and mouth movements, and are able to activate the wheels hidden under his feet to drive him forward, backward, left, and right.

The robot system is controlled by the therapist toolkit which connects the robot wearable sensors and the desktop computer.

8.4.4 *AIBO (Artificial Intelligence* ro*BO*t)

This is a robotic dog manufactured by Sony. *AIBO* has the capability to perform as companion and adjunct to therapy, especially for vulnerable persons. For example, elderly with dementia improved their activity and social behavior with AIBO as compared to a stuffed dog (Fig. 4.8b). Furthermore, children have shown positive responses. AIBO has moveable body parts and sensors for distance, acceleration, vibration, sound and pressure monitoring. One of AIBO's features is that it can

Fig. 8.3 The CosmoBot children—socializing robot. *Source* (*left*) http://protomag.com/statics/W_09_robots_cosmobot_a_sq.jpg. (*right*) http://www.hebdoweb.com/wp-content/uploads/RobotCosmobot.jpg

locate a *pink ball* via its image sensor, walk toward the ball, kick it, and head butt it. When several AIBOs interact with humans, each robot acquires slightly different set of behaviors. AIBO robot dog is shown in Fig. 8.4a.

During the interaction with children, AIBO offers to them its paw, and can respond with pleasure (green light) or displeasure (red light) after some kinds of interactions. Another dog-like robot similar in size and shape to AIBO is the robot *Kasha* (Fig. 8.4b). Kasha can walk, make noise, and wag its tail. However Kasha has not the ability to respond physically or socially with its environment, like AIBO.

In [12], the AIBO robotic dog was used in experiments (using a repertory of online questions) aiming at investigating people's relationships with it. The term "*robotic others*" has been proposed for the "*socialized robots*". The new concept is not based on human-like social interactions with the robotic pet, but it allows many criteria that compromise its "*otherness*". Humans don't simply follow social rules

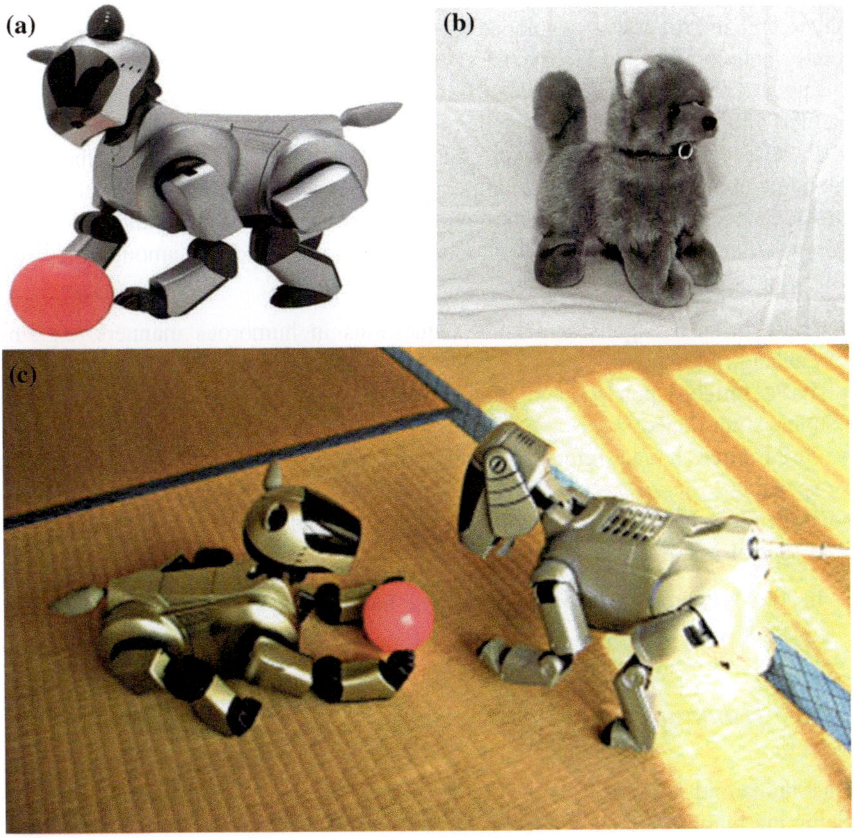

Fig. 8.4 **a** AIBO robot dog playing with the pink ball ('aibo' means partner in Japanese), **b** The robot toy Kasha [25], **c** Two AIBO's playing with the ball. *Source* (**a**) http://www.eltiradero.net/wp-content/uploads/2009/09/aibo-sony-04.jpg, (**c**) http://www.about-robots.com/images/aibo-dog-robot.jpg

but modify them. The term "robotic others" embeds robotic performance within a rich framework which is fundamentally engaged in the relationship of human and human-other. In [12], the psychological impacts of interacting with *robotic others* are explored through four studies (preschool children, older children/adolescent, longer-term impact on health and life satisfaction, online forum discussions). These studies attempted to provide some measure of social companionship and emotional satisfaction. In comparison with Kasha the children (autistic 5–8 year old) spoke more words to AIBO, and more often engaged in verbal interaction, reciprocal interaction, and authentic interaction with AIBO, typical of children without autism. More details on these studies are given in Sect. 8.6.1.

8.4.5 PaPeRo (Partner-type Personal Robot)

This socialized robot was developed by the Japanese company '*NEC Corporation*'. It has a cute looking and has two CCD cameras which allow it to see, recognize and distinguish between various human faces [13]. Its development was aimed to be a partner with humans and live together with them. PaPeRo is shown in Fig. 8.5.

PaPeRo can communicate with other PaPeRos or with electrical appliances in the home operating and controlling their use in place of humans. The PaPeRo robot was advanced further for interaction with children. This version is called *Childcare Robot PaPeRo* [14]. The interaction modes of this robot include, among others, the following:

- **PaPe talk** It responses to child's questions in humorous manners, e.g., by dancing, joking, etc.
- **PaPe touch** Upon touching its head or stomach PaPe touch will start an interesting dance.
- **PaPe Face** It can remember faces and identify or distinguish the person speaking to it.
- **PaPe Quiz** It can give a quiz to children and can judge if their answer, given by a special microphone, is correct.

In 2009 NEC released *PaPeRo Mini* with half the size and weight of the original.

8.4.6 Humanoid Sociable Robots

Four humanoid sociable robots capable of entertaining, companionship and servicing the humans are shown in Fig. 8.6.

Another humanoid socialized robot is *KASPAR* (see Fig. 4.31) which was designed for interaction with children suffering by *autism*. Autism is a disorder of neural development characterized by impaired social interaction and communication, and restricted and repetitive behavior. Autism starts before the child is three

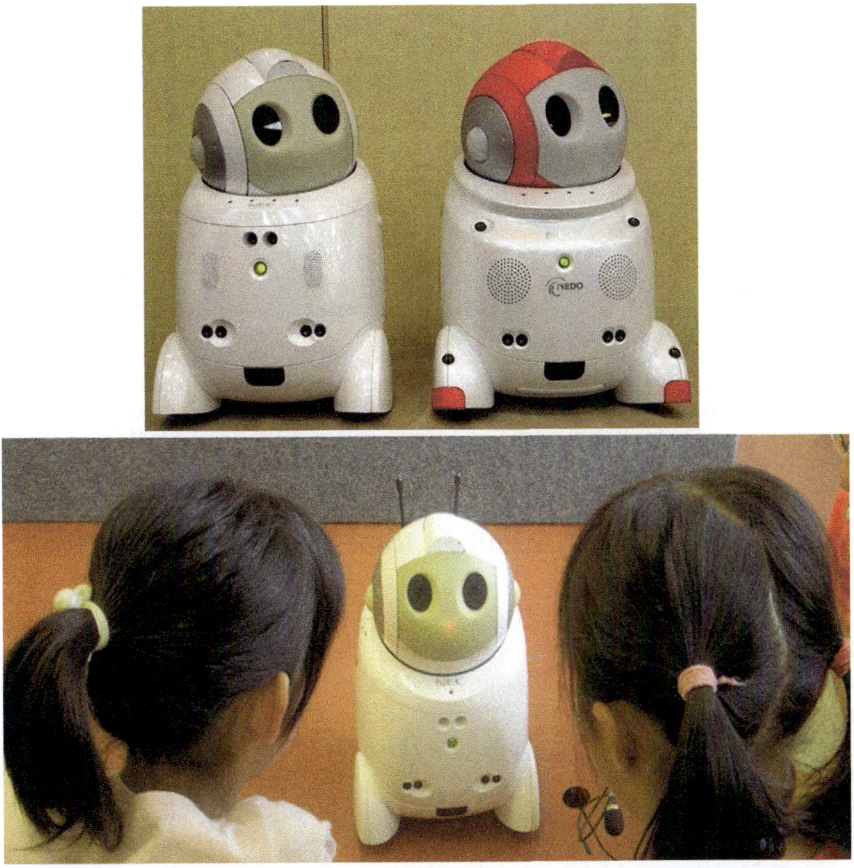

Fig. 8.5 The PaPeRo robot. *Source* (*Up*) http://p2.storage.canalblog.com/25/34/195148/8283275.jpg. (*Down*) http://www.wired.com/images_blogs/wiredscience/images/2008/12/18/papero.jpg

years old. Children with strong autism suffer from more intense and frequent loneliness than normal children. This does not imply that these children prefer to be alone. People with autism do not typically have the ability to speak naturally so as to meet their everyday communication. KASPAR belongs to the class of socialized robots that are used for autism therapy following the "*play therapy*" concept, which helps in improving quality of life, learning skills and social inclusion. According to the *National Autistic Society* (NAS; www.nas.org.uk), "play allows children to learn and practice new skills in safe and supportive environments" [5, 15]. Socialized robots for autistic children therapy are more effective and joyful than computers which were in use for long time. Actually, developing children can experience social interaction as rewarding to them. It was shown in several studies that besides KASPAR, the robots AIBO, PARO, PaPeRo, etc., are very competent in the therapy of children with autism. A comprehensive evaluation of the use of robots in autism therapy is given in [16].

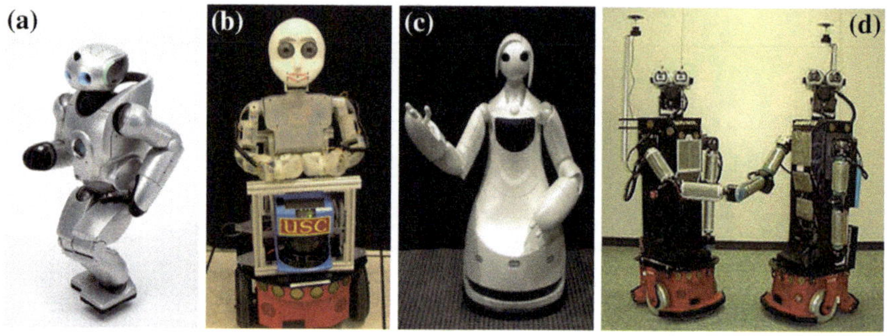

Fig. 8.6 Humanoid robots **a** Qrio entertainment robot, **b** USC Bandit robot, **c** hospital service robot, **d** the Robovie sociable robot. *Source* (**a**) http://www.asimo.pl/image/galerie/qrio/robot_qrio_17.jpg, (**b**) http://robotics.usc.edu/~agents/research/projects/personality_empathy/images/BanditRobot.jpg, (**c**) http://media-cache-ec0.pinimg.com/736x/82/16/4d/82164d21ec0ca6d01cffafbb58e0efc5.jpg, (**d**) http://www.irc.atr.jp/~kanda/image/robovie3.jpg

8.5 Ethical Issues of Socialized Robots

Socialized robots are produced for use in a variety of environments that include hospitals, private homes, schools, and elderly centers. Therefore, these robots, although intended for users who are PwSN, they have to operate in actual environments that include family members, care givers, and medical therapists. Typically, a socialized robot is designed such that it does not apply any physical force on the user, although the user can touch it, often as part of the therapy. But in most systems no physical user-robot contact is involved, and frequently the robot is not even within the user's reach. However, in the majority of cases the robot lies within the user's *social interaction domain* in which a one-to-one interaction occurs via speech, gesture, and body motion.

Therefore, the use of socialized robots raises a number of ethical concerns belonging to the psychological, social, and emotional sphere. Of course, the medical ethics principles of beneficence, non-maleficence, autonomy, justice, dignity, and truthfulness are also applied here in the same way as in the case of all assistive robots. Socialized robots can be regarded as a class of assistive robots, and indeed many researchers investigate both types under the same heading: *socially assistive robots* or *service robots* (also including household, domestic, and municipal service robots) [17, 18]. Domestic and municipal robots do not impose any ethical dilemmas different than those of industrial robots. Therefore, given that the class of assistive robots that may exert forces to the users was studied in the preceding chapter, here we will focus our investigation to the ethical implications of socialized robots. The two primary groups of socialized robots' use are: *children* and *elderly*. But, actually, there is still a need for the development of uniform ethical guidelines, control and utilization of robots in caring for children and elderly, as has been argued, e.g., by *English roboticists* [19]. Although, many guides on standards for industrial robots in the factory floor exist (e.g., ISO10218-1: 2006), these standards are not applicable to service/assistive robots, and so they have to be properly extended. A notable example towards this direction is provided in a European Union project report [20]. Many studies have revealed that healthy persons show a limited trust in autonomous assistive robots for personal care, while a number of roboticists are still expressing their concerns over the potential implications of using some of assistive robot types into human society. Of course, there are clear differences among various countries. What may be acceptable in one society (culture), e.g., Japan or Korea may look fully unacceptable in another culture (e.g., Europe).

Fundamental social and emotional (non physical) issues that have to be addressed when using socialized robots include the following [17]:

- **Attachment** The ethical issue here arises when a user is emotionally attached to the robot. Attachment can appear in all kinds of users (children, adult, elderly), and can create problems, e.g., when the robot is removed due to operational degradation or failures. The robot's absence may produce distress and/or loss of therapeutic benefits. This consequence can occur especially in users who cannot understand the reason of the robot removal. For example, it has been found that people with dementia miss the robots when they are removed. This is due to that the users feel the robots as persons. *Personification* may not be intentional but may occur when the therapist or family refer to the robot as a person (him/her) ascribing feeling to it. Here, care must be taken because patients (and humans in general) form quickly mental models of the robot as they do for humans. Of course these models do not accurately represent the real world, because robots are not humans (they have only a few human capabilities).
- **Deception** This risk can be created by the use of robots in assistive settings, especially in robot companions, teachers or coaches. The robot is typically designed to physically mimic and perform like a human when acting in these roles. Deception may also occur when the robot mimics the behavior of pets (and also using toys). It should be noted here that robots with size similar to that of users can be frightening in comparison with smaller robots. *Robotic deception* (occurring, e.g., when the patient perceives it as a doctor or nurse) may be harmful because the patient may believe that the robot can help him/her like a human (which is not true).
- **Awareness** This issue concerns both users and care givers. They both need to be accurately informed on the risks and hazards associated with the use of robots. The potential harm can be minimized by describing, as much as possible, the robot's capabilities and limitations to patients and care givers as guidelines, possibly formalized as regulations. Marketed robots are already covered by consumer protection legislation which includes instructions, warning about undesired effects, and a duty of beneficial care to the user. These regulations also hold for socialized and other service or assistive robots (see Sect. 10.6).
- **Robot authority** A robot designed to play the role of a therapist is given some authority to exert influence on the patient. Therefore the ethical question arises who actually controls the type, the level, and the duration of interaction. For

example, if a patient wants to stop an exercise, due to stress or pain, a human therapist would accept this on the basis of his general humanized evaluation of the patient's physical state. Such a feature is desirable to be technically embedded to the robot in order to balance ethically the patient's autonomy with the robot's authority.

- **Privacy** Securing privacy during human-robot interaction is of utmost importance. Patients seeking medical and rehabilitation care expect to receive respect of their privacy (this is typically backed by legislation). Robots may not have the ability to sufficiently discriminate information that can be distributed from information that should not be distributed (e.g., sensitive personal data). A robot may also not be able to distinguish between authorized and non-authorized persons to get patient's delicate information. Therefore, the patient has the ethical and legal right to be properly informed on the robot's abilities, including visual ones via cameras mounted on the robot, and transmission of acquired images to other agents.
- **Autonomy** A person mentally healthy has the right to make informed decisions about his/her care. If he/she has cognition problems, autonomy is passed to the person legally and ethically responsible for the patient's therapy. Therefore a patient (or the person responsible for his/her treatment) should be fully and reliably informed about the capabilities of the assistive/socialized robot to be used. The care giver has the ethical responsibility for this. For example, if a user is told that a robot will perform "like a pet" and later realized that this is not so he/she may be disappointed and feel lonely.
- **Human-human relation** HHR is a very important ethical issue that has to be considered when using assistive/socialized robots. Typically, robots are used as means of addition or enhancement of the therapy given by caregivers, not as a replacement of them. Thus, the patient-care giver relation and interaction must not be disturbed. But, if the robot is used as a replacement (proxy) of the human therapist, then the robot may lead to a reduction of the amount of "human-human" contact. This is a serious ethical problem if the robot is employed as the only therapeutic aid for the patient's life. In this case, in fragile persons (children with development disorders, elderly with dementia, etc.) that suffer from loneliness, the isolation syndrome may be worsened. In some cases however the use of socially assistive robots may increase the amount of patient-therapist interaction [17, 21].
- **Justice and responsibility** Here, the standard ethical issues of "fair distribution of scarce resources" and "responsibility assignment" should be addressed. The sophisticated assistive robots are usually costly, and so always the question: "are the benefits of the assistive robot worthy for the costs?" should be addressed. To answer this question, the conventional medical cost/benefit assessment methods can be used. The issue of the responsibility refers to the question: "who is responsible in case of harm?" If the cause of harm or injury is the robot malfunction the problem might be in the design, hardware or software of the robot. In this case the responsibility belongs to the designer, manufacturer,

programmer or distributor. If the cause of the harm is the user, this may happen due to the user's self-imposed error, unsatisfactory training, or over expectations from the robot. Here, concepts and issues similar to those of the surgery scenario presented in Sect. 6.4 may be considered.

8.6 Case Studies

Several studies have revealed the existence of fundamental differences between elder and younger people on cognitive and behavioral characteristics towards anthropomorphized or animalized artifacts such as humanoid robotic toys and dolls or robotic cats and dogs. To give some light on the above issues four case studies are briefly reviewed here, concerning autistic children and dementia elder people.

8.6.1 Children–AIBO Interactions

In a series of studies, e.g., [12, 22–26] the social and moral interactions of children with the robot dog AIBO (Fig. 8.4a), the non-robot (stuffed) dog Kasha (Fig. 8.4b), and a living dog has been extensively investigated. These studies included the following:

- **Preschool study** Observation of 80 preschoolers (3–5 years-old) and interviews with them during a 40-minute play period with AIBO and Kasha.
- **Developmental study** Observations of (and interviews with) 72 school-age children (7–15 years-old) playing with AIBO and an unfamiliar, but friendly, real dog.
- **Children with autism versus normal children study** Observations of the interaction of children with autism and children without autism with AIBO and Kasha, and coding their behavior.

An Internet discussion forum study was also conducted involving 6438 adult AIBO owners which analyzed and classified their posting responses. The aim of these studies was to create a conceptual framework for better understanding human-animal interaction by comparing and contrasting it to human-robotic animal interaction.

The following aspects were addressed:

- Understanding of AIBO by children (and adults) in terms of its biology.
- Understanding of AIBO by children (and adults) in terms of its moral standing.
- How these understandings about AIBO differ from those toward a living dog.
- How children with or without autism interact with AIBO and the toy dog Kasha.
- Why robotic animals can benefit children with autism.

These issues belong to the four overlapping, but non redundant, domains of human conceptualization of a robotic pet, namely:

- Biological
- Mental
- Social
- Moral

The above domains provide the basic cognitive models for the organization of thoughts, and the influence actions and feelings.

For the biological issue the following '*yes/no*' question was posed to the children: "*Is AIBO alive?*" For the mental issue the questions asked were: "*Can AIBO feel happy?*", "*Why?*", "*How do you know?*", and "*Tell me more about that*". Among the preschoolers, 38 % replied that AIBO is "alive". Among older children the "yes" replies to the "alive question" were: 23 % for 7–9 years old, 33 % for the 10–12 years old, and only 5 % for 13–15 years old. For the AIBO moral standing issue, the following questions were asked: "Is it okay or not okay to *hit* AIBO, to *punish* AIBO for wrong doing, or to *throw AIBO away* (if you decided you didn't want AIBO any more) The "not okay" answer was considered to "affirm moral standing".

The majority of the preschoolers said that it is not okay to hit AIBO, to punish AIBO, or throw AIBO away, and 78 % of them supported their replies by moral justifications on AIBO's physical welfare (e.g., "because he will be hurt") or psychological welfare ("because he will cry"). The great majority of the 7–15 years-old children was strongly against hitting or throwing AIBO away, but about 50 % of them replied that it is "okay" to punish AIBO. Over 90 % of them supported one or more of their yes/no replies with moral arguments.

In the Internet study 12 % of the replies stated that AIBO had moral standing (and rights), and should have moral responsibility or be blameworthy. Furthermore, 75 % of the members affirmed that AIBO is an artifact, 48 % that it is a life-like dog, 60 % that it has cognition/mental states, and 95 % that it is a social being. It follows from the above numbers that these AIBO owners treated it "*as if*" it were a social companion and a biological being with thoughts and feelings.

In the study of children interaction with a living dog, the 3–5 years-old children did not interact with it although they affirmed that stuffed dog's biology, mentality, social companion ship and moral standing was similar to AIBO. The 7–15 years-old children interacted with both AIBO and with one of two unfamiliar, but friendly, living dogs. These children affirmed that the living dog (as compared to AIBO) was a biological, mental, social and moral being (actually 100 % of them attributed biological nature, and 83 % moral standing).

In overall, the biological standing, mental life, sociability, and moral standing of the living dog were affirmed by children uniformly, while at the same time the children attributed these features for the robotic dog as well. Social companionship was affirmed almost equally for both the robotic dog and the living dog.

For the study of autistic children, eleven 5–8 years-old children with formal diagnosis of autism were recruited. These children had some verbal ability, without significant vision, hearing or motor impairments. Each child conducted an individual 30 min interactive session with both AIBO and Kasha in a large room. The structure of each session was the following:

- **Interaction with artifact** Two interaction modes were observed, namely: "*authentic interaction*" (touching, talking, offering or kicking a ball, gesturing to, etc.), and "*non-interaction*". Up to 5 s no-interaction was still regarded as part of the previous interaction period. After the 5+ seconds break of interaction, the non-interaction period started. Figure 8.7 provides a snapshot of the "authentic interaction" period.
- **Spoken communication to artifact** The number of meaningful words to the artificial dog was recorded.
- **Behavior interaction of normal (non-autistic) children with artifact** Five interaction behaviors were considered, namely: *verbal engagement*, *affection* (petting, touching, kissing, etc.), *animating artifact* (moving the artifact's body or part of it, help AIBO to walk or eat a biscuit, etc.), *reciprocal interaction*,

Fig. 8.7 Snapshot of authentic interaction. *Source* www.dogster.com/files/aibo-01.jpg

e.g., monitoring with hands or fingers to give a direction, verbal cues, and offering a ball or biscuit (child-artifact, child-artifact-experimenter interaction).
- **Behavioral interaction of autistic children with the artifact** A number of behaviors were observed typical of children with autism, namely: rocks back and forth, flick fingers or hand, high-pitched noise, unintelligible sounds, repeated words, lines up, inappropriate pronouns, use third person for self, withdraws/standoffish, unreasonable fear, licks objects, smells objects, lunging/starting, and self injurious.

Results: Children without autism

- Children found AIBO more engaged than Kasha.
- Children spent about 72 % of the AIBO session actually interacting with AIBO, and 52 % of the Kasha session interacting with Kasha.
- Children spoke more words per minute to AIBO than to Kasha.

Results: Children with autism

- No significant statistical differences were found between AIBO and Kasha in the exhibition per minute of any of the considered individual autistic behaviors.
- When all behaviors were combined together, the mean number of autistic behaviors per minute with AIBO was 0.75 and with Kasha was 1.1.

The above results showed that AIBO (which is a very good representative example of robotic dogs) might help the social development of children with autism. Compared to the use of Kasha (which was not able to respond to its physical or social environment) the study showed that children with autism were more engaged in the three principal health children behaviors (verbal engagement, authentic interaction, reciprocal interaction) and spoke more words to AIBO, and, while in AIBO session, they engaged in fewer autistic behaviors.

A similar study was conducted (by the same researchers) concerning the moral accountability of humanoids in the **HINTS** (*Human Interaction with Nature and Technological Systems*) Laboratory of the University of Washington (Seattle) [27]. This study was focused on whether humans assign moral accountability to humanoid robots for harm they cause, and to the issue of how a humanoid (*Robovie*, Fig. 8.6d) is perceived by a number of interviewed persons, i.e., as an artifact or something between technological artifact and human. About 65 % of the interviewed persons attributed some level of moral accountability to Robovie, while in an interaction of people with Robovie, 92 % of them exhibited a "rich dialogue", in the sense that participants were interacting with the robot in a manner "beyond socially expected". In overall, the study has revealed that "about 75 % of the participants believed that Robovie could think, could be their friend, and could be forgiven for transgression". The study was then extended to kids engaged in a game play and other interactions with Robovie. The results were analogous to those of children-AIBO interaction. Figure 8.8 shows a child and Robovie sharing a hug during the social interactions experiment at HINTS Laboratory.

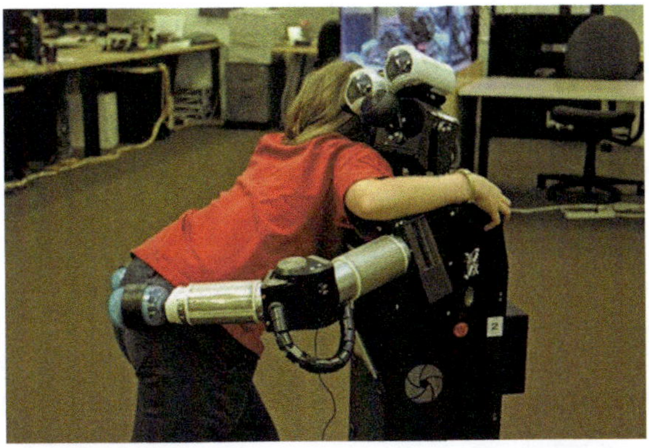

Fig. 8.8 Emotional interaction of a child with Robovie, a semi-humanoid robot developed at the Advanced Telecommunication Research (ATR) in Japan. *Source* http://www.washington.edu/news/files/2012/04/girl_robovie.jpg

8.6.2 Children–KASPAR Interactions

As indicated in Sect. 4.4.8, KASPAR is a child-sized humanoid robot which has a static body (torso, legs, hands), moving head and arms (8 DOF head, 3 DOF arms), facial expressions, and gestures. The robot communication capabilities allow it to present facial feedback, by changing the orientation of the head, moving the eyes and eye lids, and moving the arms. KASPAR can present several facial expressions which differ from each other by minimal changes around the mouth opening that also visibly affect the overall face (Fig. 8.9).

The *Adaptive Systems Research Group* at the University of Hertfordshire (U.K.), that has developed KASPAR and enhanced it over the years, has performed several experiments with autistic children [28–32]. For example, in [29] the interaction with KASPAR of three autistic children (a 6-year-old girl G, a boy B with severe autism, and a 16-years-old teenager T with severe autism) is presented. T was not tolerating any other children in play or other task-oriented interactions. In the following we give a summary of these results.

The girl G was not talking, refusing all eye contact, and in general, she was not able to interact in anyway. When KASPAR was presented to her at a distance, after some initial skepticism, she indicated her desire to go near to KASPAR (Fig. 8.10a). She then explored for some time paying her attention to KASPAR's face and eyes, and tried to imitate it when it played the tambourine. Her mother was delighted. After a little time, G stretched out her hand reaching for the experimenter's hand, something she did for the first time. In overall, KASPAR has created for G an environment where she started to play touching and gazing the thing with the experimenter's hand. This interaction was then extended and performed with the experimenter. G investigated safely several kinds of interactions with KASPAR exploring cheek stroking, and nose squeezing. Figure 8.10 shows three snapshots of G's interaction with KASPAR.

Fig. 8.9 **a** KASPAR's happiness is transferred to the girl who shows it to the teacher, **b** KASPAR's thinking expression, **c** The girl mimics KASPAR. *Source* (**a**) http://dallaslifeblog.dallasnews.com/files/import/108456-kaspar1-thumb-200x124-108455.jpg, (**b**) http://asset1.cbsistatic.com/cnwk.1d/i/tim/2010/08/25/croppedexpression.jpg. (**c**) http://i.dailymail.co.uk/i/pix/2011/03/09/article-1364585-0D85271C000005DC-940_468x308.jpg

B was a boy with severe autism that was interacting at home with other family members but at school not interacting with anybody (teachers or other children), and at the playground playing only with himself fully isolated. When KASPAR was presented to him he showed strong interest in the robot frequently exploring its surfaces by touch, and later exploring the robot eyes. He showed a fascination in KASPAR's eyes and eyelids, and at a later time he started exploring his teacher's eyes and face. Finally, after interacting with the robot once per week for a number of weeks, B started to show his excitement with his teacher, reaching out her, and asking her (nonverbally) to join in the game. This behavior was then extended to the experimenter and the other adults around him (Fig. 8.11).

As in the G's case, B showed a tactile engagement with KASPAR, mainly by touching and gazing at the robot. This engagement was then extended to co-present

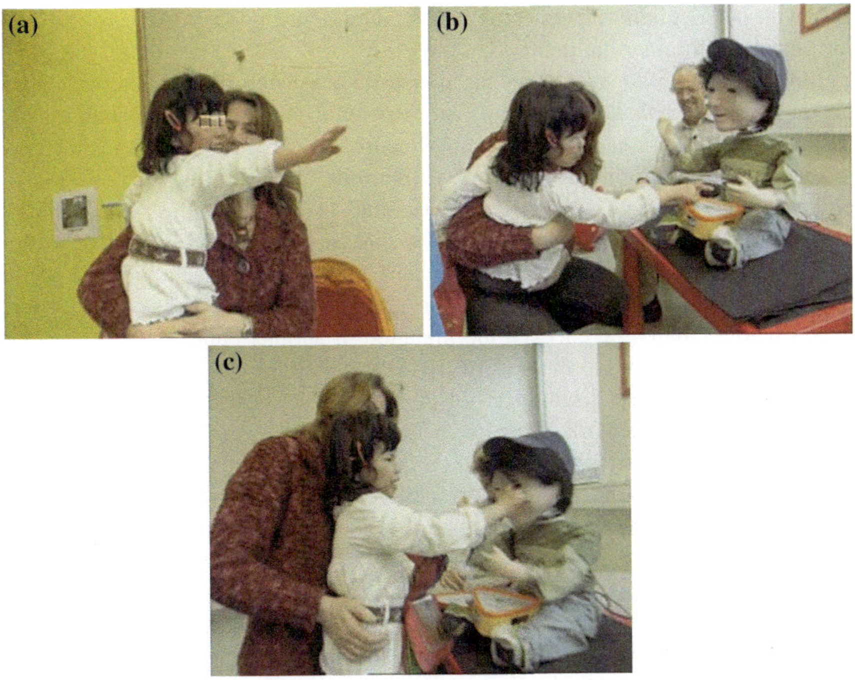

Fig. 8.10 **a** The girl G indicates her wish to move closer to KASPAR, **b** G is imitating KASPAR's drumming action, **c** G is exploring KASPAR's face and eyes. Courtesy of Kerstin Dautenhahn [29]

Fig. 8.11 **a** The boy B explores the face of Kaspar touching it, **b** B explores very closely KASPAR and then turns to his teacher and explores her face in a similar way. Courtesy of Kerstin Dautenhahn [29]

adults. KASPAR motivated B for sharing a response with his teacher, and offered an interesting, attention grabbing, interacting object which the child and teacher jointly observed in a shared way.

Teenager T When T was introduced to KASPAR he was feeling completely comfortable focusing his attention on KASPAR, exploring its facial features very carefully and exploring his own facial features at the same time. Since T refuses to play with other children, the therapy was focused on using KASPAR as a mediator for playing with other children.

Initially, T refused even with his therapist to use the robot's remote control playing only with himself. Gradually he accepted to play a simple imitation game with the therapist mediated by the robot. Finally, he learned to look at his therapist to show her how he imitates KASPAR delightfully for him. This was considered a proper starting point to introduce another child to T to play the same imitation game. T's behavior was similar to that of G and B. He moved from an exploration of KASPAR to other present adults (gazing at KASPAR's face and then at the therapist's face). Then, T was gazing at co-present others in response to the actions of KASPAR. Finally, he checked the imitation of another child.

In [30, 31], several scenarios of KASPAR-assisted play for children with physical and cognitive disabilities are presented. The experiments of [30] were conducted in the framework of the IROMEC (Interactive Robotic Social Mediators as Companions) project (www.iromec.org/) for the study of the behavior of three specific children groups, namely: (i) children with 'mild mental retardation', (ii) children with 'severe motor impairments', and (iii) children with 'autism' [33]. The socialized robot was used as a mediator that encourages the children to discover a repertory of play styles, from solitary to collaborative play with teachers, therapists, parents, etc. The therapeutic objectives adopted were selected on the basis of discussions with panels of experts from the participating institutes, and were classified according to the 'International Classification of Functioning-version for Children and Youth' (ICF-CY). The play scenarios involved three stages, viz. (i) preliminary concept of play scenarios, (ii) outline scenarios (abstract), and (iii) social play scenarios (final). These play scenarios referred to five fundamental developmental areas, i.e., sensory development, communication and interaction, cognitive development, motor development, and social/emotional development.

In [31], some of the results obtained in the framework of the ROBOSKIN project (http://blogs.herts.ac.uk/research/tag/roboskin-project) exploiting new tactile capabilities of KASPAR are described. A novel play scenario is implemented along the lines of [30]. The results of the case study examples of [31] were obtained for preschool children with autism, primary special school children with moderate learning abilities, and secondary school children with severe learning disabilities. The trials were set up such as to familiarize the children with present adults and the robot, with ultimate objective to allow them to freely interact with the robot and the adults (teachers, therapists, experimenters). The children of the preschool nursery were involved in basic 'cause effect' games, e.g., touching the side of the head to activate a 'bleep' sound, stroking the torso or leg to activate a posture of 'happiness' followed by a verbal sign such as 'this is nice', 'ha ha ha', etc. (Fig. 8.12a).

In general, the case study analysis of children responding to KASPAR's reactions to their touch, exploring 'happy' and 'sad' expressions, demonstrated that autistic children were inclined for tactile interaction with the robot, exhibiting some responsiveness to the respective reactions of KASPAR to their touch. During the initial interaction some children did not respond appropriately to KASPAR's sad expression as a result of a forceful touch, but after several session times they started

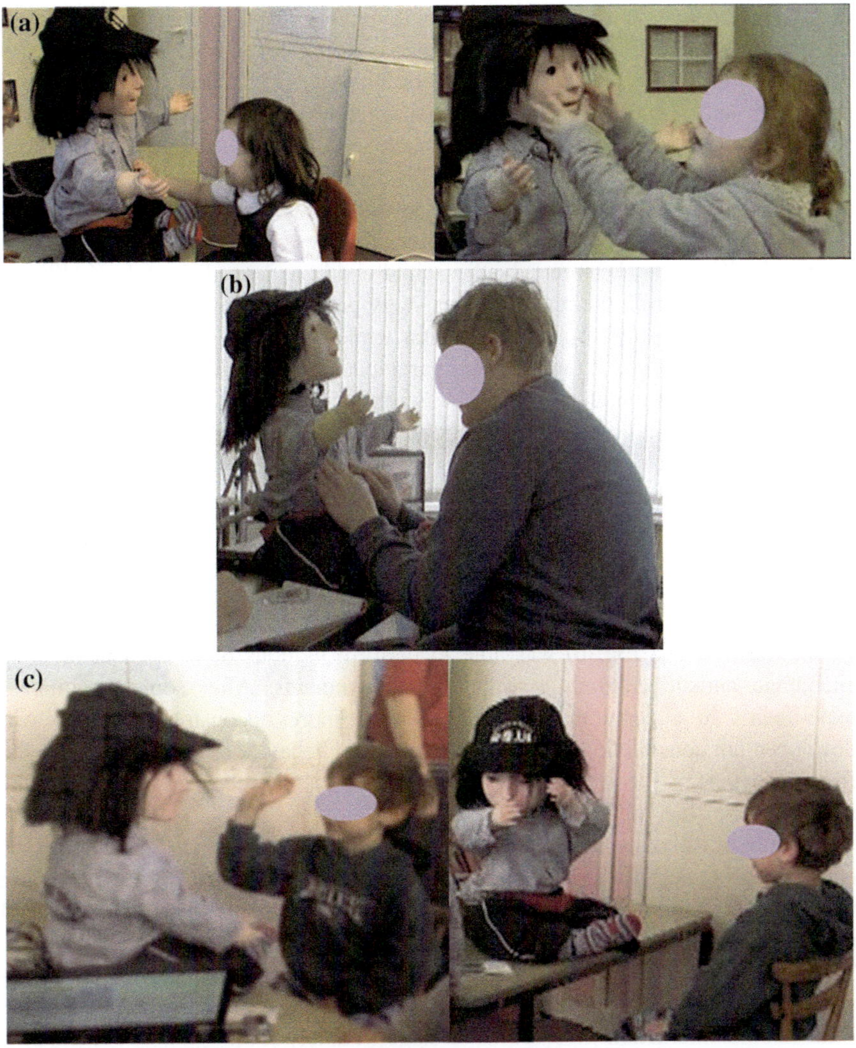

Fig. 8.12 **a** Preschool autistic children exploring tactile cause-and-effect via interaction with KASPAR, **b** A child with very low attention span learns gentle interactions (with the teacher's help), **c** KASPAR encourages or discourages certain tactile behaviors. Courtesy of Kerstin Dautenhahn [31]

to pay attention to their action, understanding (with the help of the experimenter) the cause of robot's sad expression. Then, the children started to gently stroke the robot in the torso or tickle its foot to cause it to exhibit its expressions of 'happiness' (Fig. 8.12b). Following the advice of therapists in a secondary school with very low functioning autistic children, the 'follow me' game was accompanied by verbally name and loud each body part that the robot was pointing to during the game. This proved very helpful to some children. For example, in a case it attracted the child's attention, helping him to better concentrate on the game and further developing the sense of 'self' (Fig. 8.12c). Finally, in [32] a new design, implementation, and initial evaluation of a triadic (child-KASPAR-child) game is presented. In this study each child participated in 23 controlled interactions which provided the statistical evaluation data. This data indicated that the children looked at, spoke to, and smiled more at the other autistic child, during the triadic game thus improving their social behavior. The study discusses, interprets, and explains the observed differences between dyadic (child-child) and triadic (child-KASPAR-child) play. In overall, the results of this study provided positive evidence that the social behavior of pairs of autistic children while playing an imitation-based, collaborative video game changes after participating in triadic interaction with a humanoid social robotic partner. The work described in [32] is part of the AURORA project [34] in which the robotic doll ROBOTA was also investigated as described below.

8.6.3 Robota Experiments

Robota doll (Fig. 8.13a, b) was able to perform motions of its own in order to encourage autistic children to imitate its movements. After some preliminary experiments in a constrained set-up, a much more unconstrained set-up was developed not constraining children's postures and behavior during interaction with Robota, exposing the child a number of times to the robot, also reducing the intervention of therapists.

The aim of Robota was to focus on spontaneous and self-initiated behavior of the children. Robota operates in two modes: (a) as a *dancing toy*, and (b) as *a puppet*. In the "dancing" mode the robot was moving its arms, legs and head to the beat of pre-recorded music, i.e., children rhythms, pop music, and classic music. In the "puppet" mode the experimenter was the puppeteer, and was moving the robot's arms legs or head by a simple press of buttons on his laptop. The experiment of the *child-Robota* interaction involved three stages, namely [35]:

- **Familiarization** The robot was placed in a box (black inside) similar to a puppet show. At this phase the child mostly watched while sitting on the floor or on a chair, occasionally leaving the chair to get closer to the robot and watch closely, touch, etc. (Fig. 8.13c).
- **Supervised interaction** The box was removed, the robot was placed openly on the table and the child was actively encouraged by the therapist to interact with Robota (Fig. 8.13d).
- **Unsupervised interaction** The child was not given any instructions or encouragements to interact with the robot, and was left to interact and play imitation games on his own (if wanted to do so), while the robot was operating as a puppet by the experimenter again.

Fig. 8.13 **a** The robota face, **b** the full robota doll, **c** familiarization stage, **d** supervised interaction stage. *Source* (**a**) http://www.kafee.files.wordpress.com/2008/01/robota.jpg. Courtesy of Kerstin Dautenhahn, (**b**) effects of repeated exposure of a humanoid robot on children with autism—Can we encourage basic social interaction skills? Robins et al. [35], (**c**) and (**d**) [35]

These experiments showed that allowing the autistic children to repeatedly interact with the robot over a long period of time, helped them to explore the interaction space of robot-human (and human-human) interaction. In some cases the children used the robot as a mediator, an object of shared attention, for their interaction with the experimenter. When the children became accustomed to the robot on their own, they all opened themselves up to incorporate the experimenter in their world, interacting with him/her and actively seeking to share their experience with him/her as well as with the therapist.

Some general conclusions drawn from the above and other studies conducted with autistic children are the following:

- High functioning or low functioning autistic children respond differently to the same robotic toy.
- Robotic toys can act as mediators between controlled, repetitive activities and human social interactions.
- Robotic toys able to express emotions and facial expressions can show how to recognize these cues.
- Robotic toys made of skin-like materials promote the sense of touch appeal to autistic children.
- Interactive robotic toys can reinforce lessons taught in speech and music therapy classes.

In general, the question remains whether it is ethically correct to encourage children with autism to engage in affective interactions with machines that are not able to show emotional behaviors. But, from the perspective of the autistic individual and his/her needs, it is not clear that such ethical concerns are really relevant [28].

8.6.4 Elderly–Paro Interactions

Paro baby seal robot was presented in Sect. 8.4 and has been designed and constructed primarily for social interaction with elderly people, although it can also be used in infants' interactions. Most people seem to perceive the robot as an animate object and treat it as though it is alive and loving attention. Paro can make animal sounds, emotional expressions, and has the capability to learn voices and repeat users-cause emotional therapeutic responses from patients.

In a scientific study performed at the Kimura Clinic and Brain Functions Laboratory of the Japan National Institute of Advanced Industrial Science and Technology (AIST: Japan) [36] it was shown that therapy using Paro can lead to prevention of cognition disorders and contribute to improvements in long-term care. A number of elderly people with cognition disorders were asked to interact with Paro, and their brain waves were measured before and after the interaction for analysis. It was found that 50 % of the individuals participated in the study experienced an improvement in their brain function. Persons who expressed a positive attitude towards Paro have shown stronger response to their therapy with improvement in their conditions and quality of life. It was also found that Paro can decrease the need for long-term therapy.

According to the Paro web site (www.parorobots.com), the Paro robot can be used in place of animal therapy with the same documented benefits in patients treated in hospitals and extended care facilities. Paro does not present the problems occurring with live animals (it cannot bite patients, does not eat food and does not create waste). Many medical and health care practitioners have authoritatively declared that the beneficial features of Paro (e.g., spontaneous and unexpected behavior similar to that of living creatures, etc.), make Paro a lovely companion. However, some professionals believe that robotic companions (such as Paro) are very popular because people are deceived about the real nature of their relationship to the artifacts [37]. But this is not always true! Many people who clearly know that Paro is a robot choose the robotic seal. A patient in a Care Home said: "I know that this is not an animal, but it brings out natural feeling", and named the robot

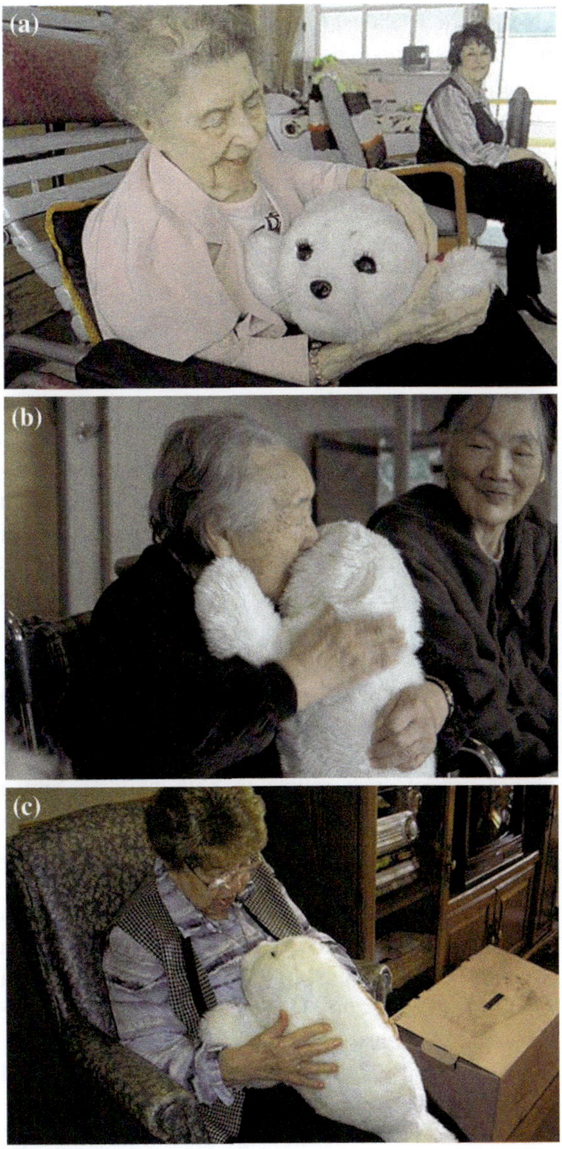

Fig. 8.14 Three elderly persons emotionally interacting with Paro. *Source* (**a**) http://www.homehelpersphilly.com/Portals/34562/images/fong_paro_north_500.jpg. (**b**) http://www.corvusconsulting.ca/files/paro-hug.png, (**c**) http://www.shoppingblog.com/pics/paro_seal_robot.jpg

"*Fluffy*". Another patient confirmed that she knows Paro is not a real animal, but she still loves it. These patients used to walk through the building halls talking to Paro as though it were a live animal.

At Pittsburg *Vincentian Home* and other elderly houses many people with dementia have experienced efficient therapy using Paro. They were typically calmed by Paro and perceived it to love them, an emotion which they actively reciprocated [38, 39]. Figure 8.14 shows snapshots of old dementia persons in emotional interaction with Paro.

Paro was surely designed not to be a replacement for social interaction with elderly, but a catalyst for initiation of social interaction. Elder people deserve proper psychological and social assistance so that they may continue finding life stimulating. Elderly do not need only medical care and food. Paro is costly, but if a decision is made to deceive dementia patients with a robot that stimulates their life and benefits them, the justification of such deception must outweigh significantly any incurrent costs to him or her. Especially, when *animal therapy* is suggested and the use of animals is prohibited for medical/health reasons then a robotic animal provides a good and low-risk alternative.

8.7 Concluding Remarks

Socially assistive robots contribute significantly in the social development of children with cognitive impairments and the improvement of the lives of elderly dementia patients. The above beneficial effects were revealed and documented in a large number of studies and individual cases all over the world.

A large repertory of humanoid and pet-like robots was already been developed and are available commercially, and many others, expectedly more sophisticated, are under development.

This chapter has outlined the fundamental issues (definitions, categories, characteristics, capabilities) of socialized/socially-assistive robots including representative well-developed examples drawn from the international scene. The chapter also included selected case studies that have documented the beneficial effect of proper use of such robots in both children and elderly with cognitive/mental impairments.

From a moral point of view these robots should only be used as mediators for therapeutical social/behavioral interactions and not as replacements of actual human care. For example, if the robot Paro is used by care personnel without properly regulating the use, or in isolation (i.e., not in groups as part of other activities) then it could be detrimental. But, if Paro is used to catalyze communication of individuals who have common interest in Paro (or have emotions of love for it) which are returned, then it can help in the improvement of their quality of life.

Of course it is difficult to foresee how much the frequency and type of human contact may be gained or lost due to the use of robot caregivers and the associated effects of this on flourishing. As indicated in [27, 40], freeing people from routine-like aspects of elderly care, will allow them to devote more of their energies to the more important duty to provide companionship and emotional support of each other. Basic in the development of companion socialized robots is the psychological understanding of emotion and interaction as well as behavior and personality, and the creation of related computational models [41–43]. Regarding the emotion aspect, several theories are available including Lazarus theory [44, 45], and Scherer theory [46].

References

1. Breazeal C (2004) Social interactions in HRI: the robot view. IEEE Trans Man Cybern Syst Part C 34(2):181–186
2. Breazeal C (2003) Towards sociable robots. Rob Auton Syst 42:167–175
3. Breazeal C (2002) Designing sociable robots. MIT Press, Cambridge, MA
4. Dautenhahn K (1998) The art of designing socially intelligent agents: science, fiction, and the human in the loop. Appl Artif Intell 12(8–9):573–617
5. Dautenhahn K (2007) Socially intelligent robots: dimensions of human-robot interaction. Philos Trans R Soc London B Biol Sci 362(1480):679–704
6. Fong T, Nourbakhsh I, Dautenhahn K (2003) A survey of socially interactive robots. Robot Auton Syst 42:143–166
7. Hollerbach JM, Jacobsen SC (1996) Anthropomorphic robots and human interactions. In: Proceedings of First Intetnational Symposium Humanoid Robotics, pp 83–91
8. Nehaniv CL, Dautenhahn K (2002) The correspondence problem. In: Dautenhahn K, Nehaniv CL (eds) Imitation in animals and artifacts. MIT Press, Cambridge, MA, pp 41–61
9. Breazeal C, Scassellati B (2002) Robots that imitate humans. Trends Cogn Sci 6(11):481–487
10. Wada K, Shibata T, Musha T, Kimura S (2008) Robot therapy for elders affected by dementia. IEEE Trans Eng Med Biol 27(4):53–60
11. Lathan C, Brisben AJ, Safos CS (2005) CosmoBot levels the playing field for disabled children. Interactions 12(2):14–16
12. Kahn PH, Freier NG Jr, Friedman B, Severson RL, Feldman EN (2004) Social and moral relationships with robotic others? In: Proceedings of 2004 IEEE international workshop on robot and human interactive communication, Kurashiki, Okayama, pp 20–22
13. Robot Pets and Toys www.robotshop.com/en/robot-pets-toys.html
14. Fujita Y, Onaka SI, Takano Y, Funada JUNI, Iwasawa T, Nishizawa T, Sato T, Osada JUNI (2005) Development of childware robot PaPeRo, Nippon Robotto Gakkai Gakujutsu Koenkai Yo Koshu (CD-ROM), pp 1–11
15. Boucher J (1999) Editorial: interventions with children with autism methods based on play. Child Lang Teach Ther **15**:1–15
16. Robins B, Dautenhahn K, Dubowski K (2005) Robots as isolators or mediators for children with autism? A cautionary tale. In: Proceeding of symposium on robot companions hard problems and open challenges in human robot interaction, Hatfield, pp 82–88, 14–15 April 2005
17. Feil-Seifer D, Mataric MJ (2011) Ethical principles for socially assistive robotics. IEEE Robot Autom Mag 18(1):24–31
18. Dogramadzi S, Virk S, Tokhi O, Harper C (2009) Service robot ethics. In: Proceedings of 12th international conference on climbing and walking robots and the support technologies for mobile machines, Istanbul, Turkey, pp 133–139
19. http://www.infoniac.com/hi-tech/robotics-expert-calls-for-robot-ethics-guidelines.html
20. http://www.respectproject.org/ethics/principles.php

21. Wada K, Shibata T, Saito T, Sakamoto K, Tanie K (2003) Psychological and social effects of one year robot assisted activity on elderly people at a health service facility for the aged. In: Proceedings of IEEE international conference on robotics and automation (ICRA), Taipei, pp 2785–2790
22. Melson GF, Kahn PH Jr, Beck A, Friedman B (2009) Robotic pets in human lives: Implications for the human-animal bond and for human relationships with personified technologies. J Soc Issues 65(3):545–567
23. Kahn Jr PH, Friedman B, Hagman J (2002) I care about him as a pal: conceptions of robotic pets in on-line AIBO discussion forums. In: Proceedings of CHI'02 on human factors in computing systems, pp 632–633
24. Kahn PH, Friedman Jr B, Perez-Granados DR, Freier NG (2004) Robotic pets in the lives of preschool children. In: Proceedings of CHI'04 (extended abstracts) on human factors in computing systems, pp 1449–1452
25. Stanton CM, Kahn PH, Severson Jr RL, Ruckert JH, Gill BT (2008) Robot animals might aid in the social development of children with autism. In: Proceedings on 3rd ACM/IEEE international conference on human robot interaction, pp 271–278
26. Friedman B, Kahn PH Jr, Hagman J (2003) Hardware companions? What on-line AIBO discussion forums reveal about the human-robotic relationships. In: Proceedings of SIGCHI conference on human factors in computing systems, pp 273–290
27. Kahn PH et al (2012) Robovie moral accountability study HRI 2012.pdf. http://depts.washington.edu/hints
28. Dautenhahn K, Werry I (2004) Towards interactive robots in autism therapy: background, motivation, and challenges. Pragmat Cogn 12(1):1–35
29. Robins B, Dautenhahn K, Dickerson P (2009) From isolation to communication: a case study evaluation of robot assisted play for children with autism with a minimally expressive humanoid robot. In: Proceedings of 2nd international conference on advances in computer-human interactions (ACHI'09), Cancum, Mexico, 1–7 Feb 2009
30. Robins B, Dautenhahn K et al (2012) Scenarios of robot-assisted play for children with cognitive and physical disabilities. Interact Stud 13(2):189–234
31. Robins B, Dautenhahn K (2014) Tactile interactions with a humanoid robot: novel play scenario implementations with children with autism. Int J Social Robot 6:397–415
32. Wainer J, Robins B, Amirabdollahian F, Dautenhahn K (2014) Using the humanoid robot KASPAR to autonomously play triadic games and facilitate collaborative play among children with autism. IEEE Trans Auton Mental Dev 6(3):183–198
33. Ferrari E, Robins B, Dautenhahn K (2010) Does it work?" A framework to evaluate the effectiveness of a robotic toy for children with special needs. In: Proceedings of 19th international symposium on robot and human interactive communication, Principe di Piemonte-Viareggio, pp 100–106, 12–15 Sep 2010
34. http://www.aurora-project.com/
35. Robins B, Dautenhahn K, te Boekhorst R, Billard A (2004) Effects of repeated exposure of a humanoid robot on children with autism: can we encourage basic social interaction skills? In: Keates S, Clarkson J, Langdon J, Robinson P (eds) Designing a more inclusive world, Springer, London, pp 225–236
36. AIST: National Institute of Advanced Industrial Science and Technology (AIST), Paro found to improve brain function in patients with cognition disorders. Transactions of the AIST, 16 Sept 2005
37. Sullins P (2006) When is a robot a moral agent? Int Rev Inf Ethics 6(12):23–30
38. Calo CJ, Hunt-Bull N, Lewis L, Metzer T (2011) Ethical implications of using Paro robot with a focus on dementia patient care. In: Proceeding of 2011 AAI workshop (WS-11–12) on human-robot interaction in elder care, pp 20–24
39. Barcousky L (2010) PARO Pals: Japanese robot has brought out the best in elderly with Alzheimer's disease. Pittsburg Post-Gazette
40. Kahn P et al. Do people hold a humanoid robot morally accountable for the harm it causes? http://depts.washington.edu/hints/publications

41. Saint-Aimé S, Le-Pevedic B, Duhaut D iGrace: emotional computational model for Eml companion robot. In: Kulyukin VA (ed) Advances in human robot interaction, in-tech (Open source: www.interchopen.com)
42. Sparrow R, Sparrow L (2006) In the hands of machines? The future of aged care. Minds Mach 16:141–161
43. Borenstein J, Pearson Y (2010) Robot caregivers: harbingers or expanded freedom for all? Ethics Inf Technol 12:277–288
44. Lazarus RS (1991) Emotion and adaptation. Oxford University Press, Oxford/New York
45. Lazarus RS (2001) Relational meaning and discrete emotions. Oxford University Press, Oxford/New York
46. Sherer KR (2005) What are emotions? How can they be measured? Soc Sci Inf 44(4):695–729

Chapter 9
War Roboethics

In war, truth is the first casualty.

Aeschylus

Never think that war, no matter how necessary, nor how justified, is not a crime.

Ernest Heningway

9.1 Introduction

Military robots are receiving considerable attention by politicians who release tremendous amounts of money for related research. The ethics concerning these robots, especially lethal autonomous robotic weapons, lies at the center of roboethics. There is a strong debate on whether they are allowed to be used or not, in modern war. This is stronger than the debate about other technological systems which are always *double-edged swords* that have benefits and drawbacks, criticisms and advocates. Supporters of their use argue that military robots have substantial advantages which include the saving of the lives of soldiers and the safe clearing of seas and streets from *improvised explosive devices* (**IED**). They also argue that robot weapons can expedite war more ethically and effectively than human soldiers who, under the influence of emotions, anger, fatigue, vengeance, etc., may overreact and overstep the laws of war.

The opponents of the use of autonomous lethal robots, argue that weapon autonomy itself is the problem and not mere control of autonomous weapons could ever be satisfactory. Their central belief is that autonomous lethal robots must be completely prohibited. Particular concerns include, among others, the uncertainty in assigning responsibility in case of robot malfunctioning that leads to war law violation, the lowering of barriers to war, and the unclear legal status of autonomous war robots as, e.g., in case of converting or using a surveillance robot as a lethal robot.

S.G. Tzafestas, *Roboethics*, Intelligent Systems, Control and Automation: Science and Engineering 79, DOI 10.1007/978-3-319-21714-7_9

The purpose of this chapter is to examine the above issues at some detail. In particular, the chapter:

- Provides background material on the war concept and the ethical laws of war.
- Discusses the ethical issues of war robots.
- Presents the arguments against autonomous war robots, also including some of the opposite views.

9.2 About War

In the *Free Merriam Webster dictionary, war* is defined as: (1) "a state or period of fighting between countries or groups", (2) "a state of usually open and declared armed hostile conflict between states or nations" (3) "a period of such armed conflict".

According to the war philosopher *Carl von Clausewitz* "war is the continuation of policy by other means". In other words war is about governance using violence and not peaceful means to resolve policy which regulates life in a territory. *Clausewitz* says that was is like a duel but on a large scale. His actual definition of war is: "an act of violence intended to compel our opponent to fulfill our will". *Michael Gelven* provides a more complete definition of war as "an actual, widespread and deliberate armed conflict between political communities, motivated by a sharp disagreement over governance", i.e., war is intrinsically vast, communal (political) and violent. In other words, according to Gelven war is not just the continuation of policy by other means, but it is about the actual thing that produces policy (i.e., about governance itself). It seems to be governance by threat, although threat of war and the occurrence of mutual dislike between political communities are not sufficient indicators of war. The conflict of arms must be actual, intentional and widespread [1–3].

L.F.L, Oppenheim defined war as "… a contention between two or more States, through their armed forces, for the purposes of overpowering each other and imposing such conditions of peace as the victor pleases" (quoted in British Manual of Military Law, Part III). A definition, of an armed conflict for the purposes of regulation, is: "An armed conflict exists whenever there is a resort to armed force between States or protracted armed violence between governmental authorities and organized armed group or between such groups within State". This definition looks at armed conflict rather than war.

A war does not really start until a conscious commitment and a strong mobilization of the belligerents occurs, i.e., until the fighters intend to go to war and until they realize it using a heavy army force. War is a phenomenon that takes place between political communities (i.e., states or intended to be states in case of civil war). It is recalled here that the concept of *state* is different than the concept of *nation*. A nation is a community which shares ethnicity, language, culture, ideals/values, history, habitat, etc. The state is a more restricted concept referring to the type and a machinery of government that determines the way life is organized in a given territory. The deep cause of war is certainly due to the human drive for dominance over others. Humans have been fighting each other since the prehistoric times, and people have been discussing its rights or wrongs for almost as long. War is a bad thing and controversial social effects raise critical moral questions for any thoughtful person. The answers to these questions are attempted by *war ethics* which will be discussed in the following.

9.3 Ethics of War

The ethics of war was motivated by the fact that it is a violent bad process and should be avoided as much as possible. War is a bad thing because it results in deliberate killing or injuring people, which is a fundamentally wrong abuse of the victims' human rights. The ethics of war aims at resolving what is right or wrong, both for the individual and the *states* (or *countries*), contributing to debates on public policy (governmental and individual action), and ultimately at leading to the establishment of codes of war [4–10].

Fundamental questions that have to be addressed are:

- Is war always wrong?
- Are there situations when it might be justified?
- Will war always be part of human experience or there are actions that can be done to eliminate it?
- Is war a result of intrinsic human nature or, rather, of a changeable social activity?
- Is there a fair way to conduct war, or is it all hopeless barbarous killing?
- After the end of a war how should post-war reconstruction proceed, and who should be in charge?

The three dominating doctrines (traditions) in the "*ethics of war and peace*" are [1, 3]:

- Realism
- Pacifism
- Just war

9.3.1 Realism

Realists believe that war is an inevitable process taking place in the anarchical world system. Classical realists include:

- *Thucydides* (an ancient Greek historian who wrote the "History of the Peloponnesan war").

- *Maciavelli* (a Florentine political philosopher who wrote "The Prince" (Il Principe)).
- *Hobbes* (an English philosopher who wrote "Leviathan", stating that the state of nature was prone to "war all against all").

Modern realists include *George Kennan, Reinhold Niebuhr* and *Henry Kissinger.*

The doctrine of realism is mostly influential to political scientists, scholars and practitioners of international affairs. The central issue of realism is a suspicion about moral aspects and justice to the conduct of international relations. According to realists war must be launched only for the country's self-interest, and once it has started, a state must do anything it can to win. In war "*anything goes*", i.e., during wartime international war laws are not applicable.

Actually, there are two distinct types of realism, namely:

Descriptive realism: According to the descriptive realism principle the states cannot behave morally in wartime either for reasons of motivation or for reasons of competitive struggle. This is because states are not heartened in terms of morality and justice, but in terms of power, security, and national interest.

Prescriptive realism: According to the prescriptive realism doctrine a prudent state is obliged to act amorally in the international scene. Realists believe that if a state is too moral, other states will quite probably exploit it and act more aggressively. Therefore, to assure state security, states must be on continuous preparation for conflict through economic and military grow-up. The general war rules endorsed by prescriptive realism are of the type: "Wars should only be conducted in response to aggression", or "During war, non combatants should not be directly targeted with killing weapons". These rules are similar to those of "just war". But in prescriptive realism they are adopted for different reasons than those of "*just war*" theory. Prescriptive realism considers these rules as useful rules, whereas *just war* is based on moral rules that have to be followed.

9.3.2 Pacifism

Pacifists reject war in favor of peace. So, actually, pacifism is "*anti-warism*". Pacifism objects to killing in general and in particular, and objects to mass killing for political reasons as it commonly happens during wartime. A pacifist does not object violence which does not lead to human killing, but rejects war believing that there are no moral grounds that justify resorting to war. A pacifist believes that war is always wrong. The major criticism to pacifism is that a pacifist refuses to take ferocious measures required for defending him/her self and his/her country. Another criticism to pacifism is that, by not resisting to international aggression with effecting means, it ends up rewarding aggression and fails to protect people who need protection.

9.3.3 *Just War Theory*

This is the theory that specifies the conditions for judging if it is just to go to war, and conditions for how the war should be conducted. *Just war theory* seems to be the most influential perspective on the ethics of war and peace. The *just war theory* is fundamentally based on Christian philosophy endorsed by such notables as Augustine, Aquinas, Grotius, Suarez, etc. The founders of *just war theory* at large are recognized to be Aristotle, Plato, Cicero and Augustine.

The just war theory tries to synthesize in a balanced way the following three issues:

- Killing people is seriously wrong.
- States are obliged to defend their citizens and justice.
- Protecting innocent human life and defending important moral values sometimes requires willingness to use violence and force.

Just war theory involves three part which are known by their latin names, namely: *jus ad bellum, jus in bello,* and *jus post bellum.*

9.3.3.1 Jus ad Bellum

Jus ad Bellum specifies the conditions under which the use of military force must be justified. Political leaders who initiate wars and set their armed forces in motion are responsible for obeying *jus ad bellum* principles. If they fail in that responsibility, then they commit war crimes. The *jus ad bellum* requirements that have to be fulfilled for a resort to war to be justified are the following[1]:

1. **Just Cause** A war must be for a *just cause*. Fear with respect to a neighboring country power is not a sufficient cause. The main just cause is to put right a wrong. The country that wants to go to war must demonstrate that there is a just cause to do so (e.g., defending against attack, recapturing things taken, punishing people who have done wrong, and to correct public evil).
2. **Right Intention** The only intention allowed for a country to go to war is for the sake of its *just cause*. For launching a war it is not sufficient to have the right reason. The actual motivation behind the resort to war must also be morally correct. The only right intention allowed for going to war is to see the just cause of concern secured and consolidated. No other intention is legal (e.g., seeking power, land grab, or revenge). Ethnic hatred or genocide is ruled out.
3. **Legitimate Authority and Declaration** For a country to go to war the decision should be made by the proper authorities of that state (as specified in its constitution), and declared publicly (to its citizens and to the enemy state(s)).

[1]Most countries accept nowadays the position that international peace and security require the *United Nations Security Council* approval prior to an armed response to aggression, unless there is an imminent threat.

4. **Last Resort** War must be the last resort. A state should go to war only if it has tried every sensible non-violent alternative first.
5. **Proportionality** The war must be in proportion. Soldiers may only use force proportional to the aim they seek. They must restrain their forces to the minimal amount sufficient for achieving their goal. Weapons of mass destruction are typically seen as being out of proportion to legal ends.
6. **Chance of Success** A country may go to a war only if it can foresee that doing so will have measurable impact on the situation. This rule aims at blocking mass violence which is vain, but it is not included in the international law, because it is considered to be against small, weaker countries.

All the above criteria must each be fulfilled for a war declaration to be justified.

9.3.3.2 Jus in Bello

'Jus in Bello' refers to justice in war, i.e., to conducting a war in an ethical manner. Responsibility for *jus in bello* norms falls mainly on the shoulders of the military commanders, officers and soldiers who plan and execute the war. They have to be held responsible for violation of the *jus in bello*, international war law, principles.

According to *international war law,* a war should be conducted obeying all international laws of weapons prohibition, e.g., chemical or biological weapons,[2] and for benevolent quarantine for *prisoners of war* (**POWs**). The *international law of war* (or *international humanitarian law*) is only about 150 years-old and attempts to limit the effects of armed conflict for humanitarian purposes. The main examples are the *Geneva Conventions* the *Hague Conventions*, and the related *international protocols* (added in 1977 and 2005). The international humanitarian law is part of the body of international law that governs the relations between States. The "*guardian*" of the *Geneva Conventions* and other international treaties for the war is the International Committee of the *Red Cross*, but without been entitled to act as "police" or "judge". These functions belong to international treaties that are required to prevent and put an end to war violation, and to punish those responsible for "*war crimes*".

The fundamental principles of the *humanitarian "jus in bello" law* are the following:

1. **Discrimination** It is immoral to kill civilians, i.e., non-combatants. Soldiers are only entitled to use their non-prohibited weapons against those who are engaged in harm. Therefore, soldiers must discriminate between (unarmed) civilians who are morally immune from direct and purposeful attack, and those legitimate military, political and industrial targets involved in basic rights-violating harm. However, in all cases some collateral civilian casualties may occur. If these

[2]The use of nuclear weapons is not prohibited by the war laws, but is a "taboo" and never used after World War II.

casualties are not the outcome of deliberate aim at civilian targets, they are considered to be excusable.

2. **Proportionality** All over the duration of the war, soldiers are entitled to use only force proportional to the goal sought. Blind bombing and over-bombing (as happened in all wars after 1900 resulting in more civilian, than military, casualties) is not ethical and is not allowed.
3. **Benevolent treatment of prisoners of war** Captive enemy soldiers cease to be lethal threats to basic rights (i.e., they are "*no longer engaged in harm*"), and so they are to be provided with benevolent, not malevolent, quarantine away from battle zones, and they should be exchanged for one's own POWs, after the end of the war.
   This rule has its origin in ancient Greece, where its philosophers advocated models of life with "*human*" as their *central value* (the "*value par excellence*"). Guided by this value, *Pausany*, the 20-year old Commander-in-Chief of the Greek army in the battle of *Plataea* (479 BC) exclaimed the phrase: "*Βάρβαροι μεν, άνθρωποι δε*" ("They may be barbarians, but they are humans"), as an argument in favor of the rescue of the Persian captives. Ancient Greek politicians also set up as a goal of the state not just to protect the life of citizens, but also to motivate them towards a life of high quality.
4. **Controlled weapons** Soldiers are allowed to use controlled weapons and methods which are "*not evil in themselves*". For example, genocide, ethnic cleansing, poisonous weapons, forcing captured soldiers to fight against their own side, etc. are not allowed in *just war* theory.
5. **No retaliation** This occurs when a state A violates *jus in bello* in war in country B, and state B retaliates with its own violation of *jus in bello*, in order to force A to obey the rules. The history has shown that such retaliations do not work and actually lead to an escalation of death and an increasing destruction of war. Winning well is the best revenge.

9.3.3.3 Jus post Bellum

This refers to justice during the final stage of war, i.e., at war termination. It is intended to regulate the termination of wars and to facilitate the return to peace.

Actually, no global international law exists for *jus post bellum*. The return to peace is left to the moral laws. Some of them, not exhaustive, are the following:

- **Proportionality** The peace recovery should be reasonable and measurable, and also publicly declared.
- **Rights vindication** The basic human rights to life and liberty, and state sovereignty should be morally addressed.
- **Distinction** Leaders, soldiers and civilians should be distinguished in the defeated country that is negotiating with. Civilians must be reasonably exempted from punitive post-war measures.

- **Punishment** There are several punishments for violation of the just war rules, such as financial restitution (subject to distinction and proportionality), reformation of institutions in an aggressor state, etc. Punishments for war crimes apply equally to all sides of the war.

From the above it is clear that a war is only a *Just War* if it is both justified, and carried out in the right way. Some wars conducted for just causes have been rendered to unjust during the war because of the way they were fought. This means that it is not only the just aim of the war, but the means and methods used to fight must be in proportion to the wrong to be righted. For example, destroying with a nuclear weapon a whole enemy's city in retaliation for the invasion of an uninhabited island renders the war immoral, despite the fact that the cause of the war was just. Finally, it is remarked that some people suggest that the "*Just War Doctrine*" is by its nature immoral, while others argue that there is no ethics for war or that the doctrine is not applicable in the conditions of modern conflicts.

In case of international armed conflict, it is often hard to determine which state has violated the *United Nations Charter.* The Law of War (International Humanitarian Law) does not involve the denunciation of guilty parties as that would be bound to arouse controversy and paralyze implementation of law, since each adversary would claim to be a victim of aggression. Therefore *jus in bello must be remain independent of jus ad bellum.* Victims and their human rights should be protected no matter to which side they belong (http://lawofwar.org).

9.4 The Ethics of Robots in War

The ethical issues that stem from existing or future robots for service, therapy and education are of more immediate concern in the case of military robots, especially war/lethal robots. Although fully autonomous robots are not yet operating in war fields, the benefits and risks of the use of such lethal machines for fighting in wars are of crucial concern.

The ethical and legal rules of conducting wars using robotic weapons, in addition to convention weapons, include at minimum all the rules (principles) of war discussed in Sect. 9.3. Assuming that modern wars follow the *just war* principles, all *jus ad belum, jus in bello* and *jus post bellum* rules should be respected, but the use of semiautonomous/autonomous robots add new rules and require special considerations.

The four fundamental questions related to the use of war robots are the following:

- Firing decision
- Discrimination
- Responsibility
- Proportionality

9.4.1 Firing Decision

At present, the decision to use a robotic weapon to kill people still lies with the human operator. This is done not only because of technical reasons, but also because of the wish to ensure that the human remains "*in-the-loop*" [6].[3] The issue here is that the separation margin between human firing and autonomous firing in the battlefield is continuously decreased. As stated in [7], even if all war robots were to be supervised by humans, one may be still in dough to what extend this is actually so. On the other hand it is not always possible to avoid giving full autonomy to the robotic system. For example, according to the US Department of Defense (Office of the Assistant Secretary of Defense) combat aircrafts must be fully autonomous in order to operate efficiently [8]. This is because, some situations may occur so quickly and need such fast information processing that we would entrust the robotic systems to make critical decisions. But the law of war demands to be "*eyes on target*" either in-person or electronically and presumably in real time [9]. If human soldiers have to monitor the actions of each robot as they take place, this may restrict the effectiveness for which the robot was designed. Robots may be more accurate and efficient because they are faster and can process information better than humans.

In [10], it is predicted that as the number of robots put in operation in the battlefield increases, the number of robots may finally be more than the human soldiers. But, even if an autonomous robotic weapon is not illegal on account of its autonomy, the *just war law* requires that targeting should respect the principles of discrimination and proportionality.

9.4.2 Discrimination

Discrimination is the ethical issue that has received most attention in the use of robot weapons. As discussed in Sect. 9.3.3.2, discrimination (distinction between combatants and civilians, as well as military and civilian objects) is at the core of just war theory [10] and humanitarian laws [11, 20]. It is generally accepted that the ability to distinguish lawful from unlawful targets by robots, might vary enormously from one system to another. Some sensors, algorithms, or analytic methods might perform well; others badly. Present day robots are still far from having visual capabilities that may faithfully discriminate between lawful and unlawful targets, even in close contact encounter [12]. The conditions in which an autonomous robot will be used, namely the battlefield and operational settings, are important issues

---

[3]It is noted that the two other categories on the amount of human involvement in selecting targets and firing are: *human-on-the loop* (robots can select targets and fire under the oversight of a human who can override the robot's actions) and *human-out-of-the loop* (robots that are capable of selecting targets and firing without any human input or interaction).

both for specifying whether the system is lawful in general, and for identifying where and under what legal restrictions its use would be lawful.

It is remarked that distinguishing between lawful and unlawful targets is not a pure technical issue, but it is considerably complicated by the lack of a clear definition of what accounts as a civilian. According to the 1944 *Geneva Convention* a civilian can be defined by commonsense, and the 1977 *Protocol I* defines a civilian as any person who is not an active combatant (fighter). Of course discrimination among targets is also a difficult error-prone task for human soldiers. Therefore the ethical question here is: "ought we to hold robotic systems to a higher standard than we have yet to achieve ourselves, at least in the near future?" In [13] it is argued that autonomous lethal robots should not be used until it is fully demonstrated that the systems can precisely distinguish between a soldier and a civilian in all situations. But in [7] exactly the opposite is stated, i.e., although autonomous (unmanned) robotic weapons may sometimes make mistakes in overall they behave more ethically than human soldiers. There, it is argued that human soldiers (even if ethically trained) have higher tendency to perform wrongly in war, and find difficulty in facing justly war situations. In [9] it is also accepted that human soldiers are indeed less reliable, and provide evidence that human soldiers may perform irrationally when in fear or stress. Therefore, it is concluded there that since combat robots are affected neither by fear or stress, may act more ethically than human soldiers independently of the circumstances. Wartime atrocities have taken place throughout the human history. Therefore it will be non realistic to think they can be eliminated altogether. On the other hand armed conflicts will continue to exist. Therefore, it is stated in [9] that "to the extent that military robots can considerably reduce unethical conduct on the battlefield (greatly reducing human and political costs) there is compelling reason to pursue their development as well as to study their capacity to act ethically". In any case, "an autonomous robotic system might be deemed inadequate and unlawful in its ability to distinguish civilians from combatants in operational conditions of infantry urban warfare, but lawful in battlefield environments with few, if any, civilians present" [14]. At present, "No one seriously expects remotely-controlled or autonomous system to completely replace humans on the battlefield. Many military missions will always require humans on the ground, even if in some contexts they will operate alongside and in conjunction with increasingly automated, sometimes autonomous, systems" [14].

9.4.3 Responsibility

In all cases of using robots for industrial, medical and service tasks the responsibility assignment in case of failure is unclear and needs to consider both ethical and legislation issues. These issues are much more critical in the case of war robots which are designed to kill humans with a view to save other humans, whereas in medical robotics the robot is designed to save human lives without taking other lives. The question is to whom blame and punishment should be assigned for improper fight and unauthorized harms caused (intentionally or unintentionally) by an autonomous robot; to the designers, robot manufacturer, procurement officer, robot controller/supervisor, military commander, a state's prime minister/president, or the robot itself? [13, 15–18]. Perhaps a chain of responsibility would be a simple solution, in which case the commanding officer is finally responsible. The situation is complicated and needs to be discussed more deeply when the robot was given higher degree of autonomy, which may make it a partially or fully moral agent in future. Two problems that may be encountered in wars using robots are [9]:

- **Refusing an order** If a robot refuses a commander's order to attack a house which is known to harbor insurgent, because its sensors "see through the walls" that are many children inside and it was programmed by the rules of engagement to minimize civilian casualties, a conflict may occur. We ought to defer to the robot which may have more accurate situational awareness, or to the commander who, on the basis of the information he/she has, issued a lawful command? On the other hand if a robot refuses and order and produces more harm, who would be responsible in this case? If we give a robot the ability to refuse orders, this may be extended to human soldiers violating the basic military principle to "obey orders"?
- **Consent by soldiers to risks** There are known many cases in modern war that a malfunctioning semiautonomous or autonomous robotics weapon has killed "*friendly soldiers*". Therefore, the question arises whether the soldiers should be informed on the risks incurred when using autonomous weapons or working with dangerous items such as explosives. Does consent to risk has any meaning, if soldiers generally have not the right to refuse a work or war order?

9.4.4 *Proportionality*

The proportionality rule requires that even if a weapon meets the test of distinction, any weapon must also involve evaluation that sets the anticipated military advantage to be gained against the predicted civilian harm (civilian persons or objects). The proportionality principle requires that the harm to civilians must not be excessive relative to the expected military gain. Of course the evaluation of collateral harm is difficult for many reasons. Clearly, difficult or not, proportionality is a fundamental requirement of *just war theory* and should be respected by the design/programming of any autonomous robotic weapon.

9.5 Arguments Against Autonomous Robotic Weapons

The use in war of autonomous robotic weapons is subject to a number of objections. Three major of them are the following:

- Inability to program war laws.
- It is wrong per se to take human out of the firing-loop.
- Autonomous robotic weapons lower the barriers to war.

A short discussion of them is as follows.

9.5.1 Inability to Program War Laws

Programming the laws of war is a very difficult and challenging task for the present and the future. The aim of this effort is to achieve autonomous robotic systems that can make decisions within the war law, respecting proportionality and discrimination, better than humans. The objection here is that quite possibly fully autonomous weapons will never achieve the ability to meet the war ethical and legislation standards. They will never be able to pass the war moral "Turing test". As discussed earlier in the book, artificial intelligence over-promised, and, as many eminent workers in the field have warned, no machine will be able through its programming to replace the key elements of human emotion, compassion, and the ability to understand humans. Therefore, adequate protection of civilians in armed conflict can be ensured only if human oversights robotic weapons.

9.5.2 Human Out of the Firing Loop

The second objection to the use of autonomous robotic weapons is that a machine, no matter how intelligent is, cannot completely replace the presence of a human agent who possesses conscience and the faculty of moral judgment. Therefore, the application of lethal violence should in no circumstances ever to be delegated completely to a machine. In [14], it is stated that "this is a difficult argument to address, since it stops with a moral principle that one either accepts or does not accept". The view proposed in [17] as "*deontologically correct*" is that any weapon which is designed to select and fire at targets autonomously, should have the capability to meet the fundamental requirements of the laws of war". In [7], it is argued that such capabilities can be achieved by an "*ethical governor*". The ethical governor is a complex process that would essentially require robots to follow two steps. First, a fully autonomous robotic weapon must evaluate the information it senses and determine whether an attack is prohibited under international humanitarian law and the rules of engagement. If an attack violates the requirement of distinguishing between combatant and noncombatant, it cannot go forward. If it does not violate, it can only fire under operational orders (human-on-the-loop). Then, the autonomous robot must assess the attack under the proportionality test. The ethical governor evaluates the possibility of damage to civilians or civilian objects, based on technical data, following a "utilitarian path". The robot can fire only if it finds that the attack "satisfies all ethical constraints and minimizes collateral damage in relation to the military necessity of the target". In [7] it is concluded that with the ethical governor, fully autonomous weapons would be able to comply with international law of war better than humans.

9.5.3 Lower Barriers to War

The third objection is that the long-run development of autonomous robotic weapons, which remove human soldiers from the risk and reduce harm to civilians through greater precision, diminishes the disincentive to resort to war. The two features of precision and remoteness, in combination, which make the war less harmful are the same two features that make it easier to undertake [13]. Politicians who feel to have a moral duty to protect lives of their soldiers may favor efforts to replace human fighters with robots. *Push-button* or *risk-free* wars that result in damaged metal instead of casualties (at least to the country that uses robots) may lessen the emotional impact that wars have currently on the people of that country. The fear is that this may make easier for a country to resort to a war. These wars may also last for longer periods of time [9]. But as argued in [6], people are almost always averse to starting an unjust war, no matter if it would lead or not to human fatalities. The fact that the war is risk-free does not by itself make it more acceptable [6]. This point of view is challenged in [9] since, as stated, "it might lead to even more *dangerously foolish* ideas, such as the idea of trying to prevent wars to be resorted by increasing the brutality of the fighting". Furthermore, risk-free wars might increase terrorism, as the only way to strike back at a country that uses only robots in war is to attack its citizens. The side that has not the technology of war robots may advocate terrorism as a morally acceptable means to counterattack. Some other concerns about robotic weapons are discussed in [19].

Taking into account the threats that fully autonomous weapons would pose to civilians (described in [20]) *Human Rights Watch* (**HRW**) and the *International Human Rights Clinic* (**IHRC**) of the Harvard Law School, made the following recommendations to all States and roboticists and other scientists working in the development of robotic weapons:

«**To All States**

- Prohibit the development, production, and use of fully autonomous weapons through an international legally binding instrument.
- Adopt national laws and policies to prohibit the development, production, and use of fully autonomous weapons.
- Commence reviews of technologies and components that could lead to fully autonomous weapons. These reviews should take place at the very beginning of the development process and continue throughout the development and testing phases».

To Roboticists and Others Involved in the Development of Robotic Weapons

- Establish a professional code of conduct governing the research and development of autonomous robotic weapons, especially those capable of becoming fully autonomous, in order to ensure that legal and ethical concerns about their use in armed conflict are adequately considered at all stages of technological development.

9.6 Concluding Remarks

War between countries and states is a phenomenon that cannot be eliminated completely but can be reasonably regulated to have lower number of victims and less material destruction. The principal approaches to war launching, conducting, and ending range from *anti-warism (pacifism)* to *realism* (war is inevitable, in war "anything goes") with intermediate the *just war* theory. Not all wars in history including recent ones have obeyed *just war* laws. In our times large-scale wars are controlled by the *United Nations Security Council*.

In this chapter we overviewed the ethical issues of using robots (semi-autonomous/autonomous) in war. To this end, a general outline of "what is war" and "what is the international war law/humanitarian law was given as a background for the discussion of the "war roboethics". Robot-based wars are actually wars that are conducted by sophisticated machines, systems and methods. Therefore, in addition to the standard ethical issues of war, more issues and concerns arise which have not yet been completely and precisely addressed. Given the gradual/incremental evolution of sophisticated robots, with more intelligence and more autonomy, further research is needed to resolve new ethical issues that are expected to arise.

References

1. Coates AJ (1997) The ethics of war. University of Manchester Press, Manchester
2. Holmes R (1989) On war and morality. Princeton University Press, Princeton
3. Stanford encyclopedia of philosophy, war, 28 July 2005. http://plato.stanford.edu/entries/war
4. BBC, Ethics of war. http://www.bbc.co.uk/ethics/war/overview/introduction.shtm (Now archived)
5. Singer P (2009) Wired for war: the robotics revolution and conflict in the 21st century. Penguin Press, New York
6. Asaro P (2008) How just could a robot war be?. IOS Press, Amsterdam, pp 50–64
7. Arkin RC (2000) Governing lethal behavior in autonomous robots. CRC Press, Bocan Raton
8. http://www.defence.gov/Transcripts/Transcripts.apsx?TranscriptID=1108
9. Lin P, Bekey G, Abney K (2009) Robots in war: issues of risk and ethics. In: Capuro R, Nagenborg M (eds) Ethics and robotics. AKA Verlag Heidelberg, Heidelberg, pp 49–67

10. Walzer M (2000) Just and unjust wars: a moral argument with historical illustrations. Basic Books, New York
11. Schmidt MN (1999) The principle of discrimination in 21st century warfare. Yale Hum Rights Dev Law J 2(1):143–164
12. Sharkey N (2008) The ethical frontiers of robotics. Science 322:1800–1801
13. Sharkey N (2008) Cassandra or false prophet of doom: AI robots and war. IEEE Intell Syst 23 (24):14–17
14. Anderson K, Waxman M Law and ethics for autonomous weapon systems: why a ban won't work and how the laws of war can, laws and ethics for autonomous weapon systems. Hoover Institution, Stanford University. www.hoover.org/taskforces/national-security
15. Asaro A (2007) Robots and responsibility from a legal perspective. In: Proceedings of 2007 IEEE international conference on robotics and automation: workshop on roboethics, Rome
16. Sparrow R (2007) Killer robots. J. Appl Philos 24(1):62–77
17. Hennigan WH (2012) New drone has no pilot anywhere, so who's accountable? Los Angeles Time, 26 Jan 2012. http://www.articles.latimes.com/2012/jan26/business/la-fi-auto-drone-20120126
18. Cummings ML (2006) Automation and accountability in decision support system interface design. J. Technol Stud 32:23–31
19. Lin P (2012) Robots, ethics and war, the center for internet and society. Stanford Law School, Nov 2012. http://cyberlaw.stanford.edu/blog/2010/12/robots-ethics-war
20. HRW-IHRC (2012) Losing humanity: the case against killer robots, Human Rights Watch. (www.hrw.org)

Chapter 10
Japanese Roboethics, Intercultural, and Legislation Issues

To the city and to the individual it is advantageous to enact the common interest and not the personal.

Plato

A nation, as a society, forms a moral person, and every member of it is personally responsible for his society.

Thomas Jefferson

10.1 Introduction

The material presented in Chaps. 1 through 9 was based on the European/Western philosophy, morals, and literature on roboethics. In the present chapter an effort will be made to summarize *Japanese roboethics* on the basis of relevant results and knowledge published by native Japanese authors [1–5]. Japan is a country where even mountains, trees and pebbles are traditionally believed to have souls. Japanese people view their artifacts and tools (robotic toys, dolls, pets etc.) with affection, give to them names and treat them almost as family members. In general, Japanese culture is unique in many issues, being marked by a blend of the old and the new. Among the influx of technology and modern art, like *anime* and *manga* (comics), one can find clear and direct references to the past (such as figures of *warriors* and *gheisas*) [6]. Western media often cites *Shinto* (indigenous spirituality) as the reason for the Japanese affinity for robots. This chapter will also look at other indigenous traditions that have shaped Japan's harmonious feeling for intelligent machines and robots, particularly for humanoid ones.

South Korean people also strongly involved with robotics in anticipation of future robots such as *humanoids*; *androids* (from the Greek "άνδρας": andras = man) or *gynoids* (from the Greek "γυνή": gyni = woman). In South Korea a "*Robotics Ethics Charter*" was established which created a code of ethics for the human-robot coexistence [7, 8].

S.G. Tzafestas, *Roboethics*, Intelligent Systems, Control and Automation: Science and Engineering 79, DOI 10.1007/978-3-319-21714-7_10

Specifically, the purpose of this chapter is the as follows:

- To provide general overview of Japanese indigenous ethics and culture.
- To provide a discussion of the basic aspects of Japanese roboethics and differences from Western roboethics.
- To discuss some fundamental issues of intercultural philosophy.
- To outline the basic intercultural issues of infoethics and roboethics (aim, interpretations, shared norms, and shared values).
- To give a brief description of the legislation about robots in Europe and Korea (including the Korean Robot Ethics Charta).

10.2 Japanese Ethics and Culture

Ethics reflects the mentality of people. Japanese Ethics and West ethics have substantial differences due to their religious, cultural and historical, backgrounds. Japanese ethics is based on several indigenous concepts such as *Shinto*. Japanese social tradition was evolved along two paths [1]:

- Avoidance of abstract concepts in several issues of life.
- Avoidance of straight forward emotional expression.

As a result, social life issues, such as ethical concerns, evaluation of objects and events, incidents of war periods, etc., are considered in the cultural contexts oriented towards:

- Persons-things direct bonds (*Mono*)
- Persons-persons relations (*Mono*)
- Events (*Koto*)
- Inner minds (*Kokoro*)

These are taken place via mediated indirect ways of expressions of common (shared) senses, feelings or emotions in Mono and Koto situations.

The word for "*Ethics*" *in* Japanese is "*Rinri*", which means the study of community, or how an harmony in human relations is achieved. In the West, ethics has a more individualistic or subjective basis. According to Tetsuro Watsuji (1889–1960), Rinri is the study of *Ningen* (human beings). *Ningen* comes from *Nin* (human being) and *gen* ("space" or "between"). Thus *Ningen* is the "betweeness" of individuals and society.

The word *Rinri* is composed from two words: *Rin* and *Ri*. Rin means a group of humans (community, society) that is not chaotic, i.e., it maintains an order, and Ri indicates the course (or *reasonable way*) of keeping the order. Therefore Rinri actually means the proper and reasonable way to establish the order and keep harmonious human relationships. The question rised here is what is the reasonable/proper way of achieving order. In modern Japan Rinri withholds somehow the *Samurai code*. During the period of Sixteenth to eighteenth centuries (*Edo period*) the ethics in Japan followed the ideas of *Confucianism* and *Bushi-do*, the way of Samurai warriors ensured the maintenance of its regime (absolute loyalty and willingness to for one's lord), Confucian ethics [2].

In *Rinri* the concept of social responsibility (moral accountability) for one's action exists from the classical Japanese period, in which the individual was inseparable from the society. Each person has a responsibility towards the community, and the universe that comprehends communities. This type of ethics, i.e., the dominance of social harmonization over the individual subjectivity, is a key characteristic of Japanese ethics.

10.2.1 Shinto

Shinto (or *Kami-no-michi*) is the indigenous spirituality of the people of Japan. It contains a set of practices that must be carried out diligently in order to establish a connection between modern Japan and its ancient past. Nowadays, Shinto is a term that refers to public *shrines* suited for several purposes (e.g., harvest festivals, historical memories, war memorials, etc.). Shinto in Japanese means "*the way of Gods*", and in modern literature, the term is often used with reference to *Kami worship* and related theologies, rurals and practices.

Actually, Shinto is meant to be the "*Japanese traditional religion*" as opposed to foreign religions (Christianity, Buddhism, Islam, etc.). Etymologically, the word *Shinto* is composed from two words: *Shin* (spirit) and "*to*" (philosophical path). The word *Kami* in English is defined as "*spirits*", "*essences*" or "*deities*". In Japan most "life events" are handled by Shinto, and "death" or "after-death-life" events are handled by "*Buddhism*". A birth is celebrated at a Shinto shrine, whereas a funeral follows the Buddhist tradition which emphasized practices over beliefs. Contrary to most religions, one does not need to publicly confess belief in Shinto to be a believer.

The core of Shinto involves the following beliefs:

- *Animist world* Everything in the world was created spontaneously and has its own spirit (tama).
- *Artifacts* They are not opposed to nature and can be used to improve natural beauty and bring good.

This means that Japanese people believe in the supernatural creation of the world, where all creatures (sun, moon, mountains, rivers, trees, etc.) have their own spirits or gods and are believed to control the natural and human phenomena. This belief influences the relationship of Japanese people with nature and spiritual life, and was later expanded to include artifacts (tools, robots, cars, etc.) that bring good. The spirit of an object is identified with its owner.

The main differences of West and Japanese views about ethics are:

- *West ethics* is based on the "hierarchical world order", in which humans are above animals and animals are above artifacts, and on the "coherent self" (not accepted by Buddhism).
- *Japan ethics* is based on the exploitation of the relation among human, nature, and artifacts.

10.2.2 Seken-tei

The concept "*seken tei*" is usually interpreted as "*social appearances*". Seken-tei was derived from the warrior class ("*shi*", "*bushi*" or "*samurai*") who were concerned to keep "face" the honor of their name and status among the contemporaries. Despite the strong social changes after the end of World War II, seken-tei continues to have considerable influence on the Japanese mind set and social behavior [4].

The term "seken-tei" is composed from the word "*seken*" (community in which people share daily-life, e.g., shopkeepers, physicians, teachers, neighbors, friends) and the suffix "*tei*" (which refers to "*appearances*"). Thus, "*seken-tei*" is how we appear before the people of "*seken*" (social appearances).

Pictorially, from a "*socio-psychological*" point of view the structure of 'seken' can be represented using concentric circles. The innermost circle involves the family members, relatives and intimate friendships. The outmost circle involves the strangers to whom we are indifferent. The intermediate area would be structured in subdivisions with narrower "*seken*" (colleagues, supervisors, and any one who knows you). This is a key element of Japanese social behavior. The belief that a member of the "seken" group disapproves of some action of your is very strong and can reach levels akin to mental torture, causing very extreme reactions such as a suicide.

The lives of Japanese have several contradictory attitudes. One of them concerns privacy [3]. People want to be free, and have the right to control one's personal information. However, at the same time most of them want to get "true friends" by sharing their secret information concerning their private affairs. Also, the majority of Japanese do not like the media to get into the privacy of the victims of crime. But also many Japanese think that personal information of such victims, including their occupation, human relations, personality, life history, etc. are needed in order to know the "deep reasons and meanings" of the crime.

Actually, this is due to a dichotomy between *Seken* and *Shakai* [3]. Seken consists of traditional and indigenous world views or ways of thinking and feeling. Shakai is another aspect of the world referring to *modernized* (or *westernized*) world views or paths of thinking influenced by the systemic concepts imported from the Western countries. This *dichotomy* between *Seken* and *Shakai* helps to obtain a deep insight into Japanese mentality and explains, at least partially, the contradictions mentioned above.

To understand better the apparently contradictory features of Japanese culture the Seken-Shakai dualism has been enhanced by the *Ikai* concept, such that to get the trichotomy "*Seken-Shakai-Ikai*". *Ikai* is the world of others, i.e., the hidden or forgotten meaning of values in Seken or Shakai, as normal aspect of the world. *Ikai* is the aspect that emerges from evils, crimes, disasters and impurity along with the freedom related to art and other spiritual meaning. The concept of *Ikai* (also called *Muen*) is still under investigation and plays a crucial role for the deeper understanding of the Japanese mentality, culture and society in our modern area [3].

10.2.3 Giri

A further Japanese indigenous cultural concept is *Giri* which is generally interpreted as "*duty*" or "*obligation*" that arises from a social interaction with another person. But this interpretation does not reveal a wide range of important "*tints*" [4]. The concept of "*giri*" is constituting even in present days a significant part of Japanese social relationships and has been a standard topic in various artistic "plays", "pupped dramas", "cinema movies", and "TV operas", and draws tears from the audience. *Giri* is dynamic and complex, and is raised from a mixture of *obstinacy*, consideration of others, community duty, and moral indebtedness. Giri contact is not the outcome of an agreement between the parties involved, and almost always there is fuzziness as to whether what is done is sufficient. Therefore, in many cases it leads to a sense of frustration. Actually, Giri actions are subjective and depend on the sensitivity of the affected parties. In a "*giri*", the personal considerations are not ignored or unclearly separated. In "giri", social rules are typically regarded as obstacles to a "*giri*" relationship, although these rules may be violated when this is justified by particular circumstances. Furthermore, the human behavior in relations under "giri" seems to be more humanistic rather than obeying cold rules and regulations that cannot be adequately flexible. Finally, in disputes occurring in interactions under "giri" an effort is made to act spontaneously with consent rather than to force agreement. The result of this is the occurrence of a big gap between the expectations of legal codes and the everyday reality, which is the outcome of many compromises based on human relationship considerations. Although it might be seen strange, it is true that in practice lawyers and courts do not appear to have a dominant role and are actively avoided in "giri" affairs. Therefore, in Japan "*sincerety*" (sei-i) has a greater significance than "*rights*" in any disagreements or disputes among people.

Some examples of social behavior under Giri are the following [4]:

- "*Giri*" *dealing with* "*obstinacy*" A husband (A) and his wife (B) live with A's mother C. C becomes "ill in bed" and at the same time B's mother also becomes bedridden in her bed in B's birthplace, another city. B's father is taking care of his infirm wife. A suggests to B. You had better go home to take care of your own mother, and I will take care of my mother. But B under "giri" does not follow A's proposal.

- *"Giri" dealing with "consideration of another person"* In the above example, A's proposal to his wife was made by reference to "giri" and did not reflect his real wish. Actually, he did not want his wife to leave him and go to look after her mother, but "giri" obliged him to say so.
- *"Giri" dealing with an "exchange of favors"* The persons D and E are closely cooperating in their business. To acknowledge this during the year a gift is sent by D to E's house.

To summarize, Japan's social rules are complex and require a deep understanding of etiquette and tiered society. A first sign to this is the language differences (honorific speech, viz. *keigo, sokeigo* or *kensongo*) related to talking to superiors or those in a position of power. Basic things such as avoiding eye contact and speaking in turns are considered a sign of education and politeness. The Japanese *"tea ceremony"* (cho-no-yu) is a typical clear example of social interaction where participating people are expected to know not only how to present the tea itself but also the history and social traditions surrounding the ceremony. Guests to a tea ceremony have to learn the appropriate gestures and phases to use and the proper way to drink the tea. Some cultural aspects that are unique to Japan became a special symbol all over the world. One of them is *Gheisa*, the noted entertainers who are skilled in music, poetry and traditional dance.

10.3 Japanese Roboethics

Japanese roboethics is driven by the general cultural and ethics (*Rinri*) indigenous behaviors and beliefs including *animism*, and by the Japanese modernization or westernization (*shakai*). Western and Japanese scholars have revealed interesting differences between West roboethics and Japanese roboethics. In the West, roboethics is concerned about how to apply robots to human society driven by the fear that robotics (and advanced technology) may be turned against humanity or the essence of human beings. In Japan, at any stage of development a robot (*robotto* in Japanese) is considered as a *machine* that brings *good*; hardly an *evil* entity. The tendency in Japan is to see how to relate to the real world via a dialogue among humans, nature and robots. In Japan, research and development in robotics gives emphasis on enhancing the mechanical functionality of robots, giving little attention to ethical issues coming from the use of robots. Instead, the focus is on the legal issues for the safe use of robots.

Japanese ethics (Rinri) has the dominant role to set robots in the ethical system based on the *animism* point of view. A robot is regarded to have an identity with its owner, and as far as the owner treats it (or its spirit) in the proper way the robot must respect its owner, act in harmonious manner, and in general show an ethical behavior. Actually, robots can have their identification only while their owners are using them. Spatially, the togetherness of the existences of the human (the owner) and the robot (the machine) determines the limit of their "betweeness".

Japan is a leader in robotics (sometimes called the "*kingdom of the robot*") and is giving emphasis on the development of advanced socialized robots. The Japanese "Ministry of Economy, Trade and Industry" (**METI**) intends to make the robot industry an important industrial sector with competitive advantages both locally and internationally. To this end, the "*Next Generation Robot* (**NGT**)" project was launched supported with a large amount of money. The **NGT** project aims at producing the proper technological advances, which will improve the *symbiosis* (living together) and cooperation of humans and robots in order to enhance the quality of human life.

In Japan, autonomous and intelligent robots are easily acceptable socially because of the belief in their spirit. This facilitates the preparation of practical guidelines for the functional development of robots.

Already, many Japanese socialized robots are in the international market such as those presented in Chap. 8 (Sony's QRIO, Honda's ASIMO, NEC's PaPero, AIBO pet-like robot, AIST's Paro baby seal robot, etc.).

A great deal of research effort is given to *humanoid robots* which is due to the challenging assumption that in the near future robots will work directly alongside humans. Here, important questions to be addressed are the following:

- Why humanoid robots are preferred to have an anthropomorphic form?
- Why humans are excited by making other human-like artifacts?
- What is more humanoid, a robot with anthropomorphic form and full master-slave control or a super-intelligent metallic machine (computer)?

A leading Japanese company pursuing continuous research on humanoid robots is HONDA. Figure 10.1 shows the evolution in the R&D of Honda's humanoid ASIMO.

Fig. 10.1 Evolution of Honda's humanoid robot ASIMO. *Source* http://world.honda.com/ASIMO/history/image/bnr/bnrL_history.jpg

Very often, scholars around the world state that discussions of ethical issues of robotic applications seem to be less popular in Japan. *Kitan* [5] argues that this is not true, and this apparent lack of ethical discussions in Japan is due to the cultural and industrial differences between Japan and the West. Kitano provides a theory of Japanese roboethics. He argues that: "Robotic researchers are no more trying to purely replicate nature's objects and living beings. Their goal is not to put machines in place of, or replace, human beings, but rather to create a new tool which is designed to work with humans with any form yet seen".

Galvan, a Catholic priest and philosopher, attempted to find an answer to the question: "*What is a human being*?" [5, 9]. The answer to this question involves also the answer to the questions: "What and where the line between a human being and a humanoid robot is"?

"Where are the boundaries of human kind"? "Are there some ethical boundaries to be set"? "How far can we go"? According to him, technology helps mankind to distinguish itself from the kingdom of "animals". The key distinction of humans from humanoids is *free will*, and so humanoids can never replace a human action which has its origin in free will.

In the post-war period, besides the economic and industrial growth, Japan had many popular robot characters animated in movies and TV, e.g., [5]:

- *Mighty Atom*: A hero that saves humans against the evils and represents a "child science".
- *Dora Emon*: A pet robot which is the best friend of a human boy (Nobita).

The difference from Western robot characters (such as those of Shelley and Capek), and characters (such as those of Asimov) is that all Japanese imaginations are working to save the human kind, thus presenting an *harmonious symbiosis* of robots and humans.

To face the sharp rise of elderly people in the population, Japanese government (**METI**) is cooperating with research institutes and robotics societies (such as **JARA**: Japanese Robot Association), in order to develop and put in operation a large range of service robots (house hold robots, medical/assistance robots, socialized robots, etc.).

10.4 Intercultural Philosophy

As discussed in Chap. 2, ethics has its foundation in philosophy and mindsets of people. Therefore, before examining the *intercultural roboethics* (**IRE**) and *info-ethics* (**IIE**: Intercultural Information Ethics), it will be helpful to discuss some issues of "*intercultural philosophy*".

European philosophy (more generally, *Western philosophy*) has its origin in the ancient Greek philosophy which was concerned with two principal questions, namely: "*τι εστίν*" ("ti estin" = "*what is*") and "*τι το όν*" ("ti to on" = "*what is being*") which refer to the essence of "*existence*". The question "*what is*" was coined by the pre-Socrates philosopher Parmenides, born in the Greek colony Elea at about 515 B.C. He explains that "*reality*", i.e., "*what is*", is one (change is not possible), and "*existence*" is timeless, uniform, necessary, and unchanging. His only known work is a Poem "*On Nature*" which contains three parts:

- *Προοίμιον* ("proem" = "introduction")
- *Αλήθεια* ("aletheia" = "the way of truth")
- *Δόξα* ("doxa" = "the way of appearance opinion")

His ideas influenced the entire Greek and Western philosophy. The "*way of truth*" approach discusses "*what is real*" in contrast to "*what is illusory*", resulting from the sensory capabilities. He calls this illusory idea "*the way of appearances*". According to Parmenides: "*thinking*" and the "*thought that it is*" are the same, because you will not find "*thinking*" apart from "*what is*", in relation to which it is pronounced, and because "*to be aware and to be are the same*". It is necessary to *speak* and to *think what is*, because "*being is*" but "*not being is not*". The basic aspect of this philosophy is to investigate and understand the meaning of the word "*what*" whenever we ask "what is the essence of something?". The meaning of the word "*what*" was not agreed to be the same by all philosophers. Socrates, Plato, Aristotle and other Greek philosophers were using this word giving to it different interpretations. European philosophers like Kant, Hegel and Descartes developed their philosophies using further different meanings of the word "*what*" and of the essence of "*existence*". René Descartes (1596–1660) has put the foundation for the 17th century continental *rationalism.* His philosophical approach is known as *Cartesian doubt* or *Cartesian skepticism and* questions the possibility of knowledge putting as its goal the sorting out true from false knowledge claims. His philosophical proof of "*existence*" was stated by his famous conclusion: "*Cogito ego sum*" ("Je pense donc je suis", "I think, therefore I am"). This is in agreement with Parmenides conclusion, derived in another way. The question of *Being* has also been extensively studied by *Martin Heidegger* (1889–1976) who was using an ontological, existential, phenomenological, and hermeneutics, approach. He was concerned with European philosophy, and has pointed out that the answers to these questions do not actually lead to a kind of dialectic process, but to a "free sequence". Therefore European philosophy from its origin and its further development is not confined by only Greek philosophy. European philosophy (*euro centrism)* is a *mono-cultural* philosophy and its inner dialogue is restricted to those who share this questioning within the European culture, although there is not an homogenous European cultural environment [10]. Heidegger was interested in initiating multicultural dialogues. Here his book "Dialogue on Language between Japanese and an Inquirer" is mentioned. One of his well known statements is "we don't say: *Being is. Time is,* but rather *there is Being* and *there is Time*".

To develop an intercultural philosophy, other philosophies like Indian, Chinese, Islamic, African or Latin American philosophies should be considered and integrated to the maximum extent [11]. This can be done through communication and collaboration between these traditions and cultures, especially in the present "*globalization era*", given that these intercultural interactions are facets of human existence. Nowadays, it is no longer sufficient and important to philosophize in a very regional way but in an intercultural way. Advances in transportation, communication and the Internet play an important role in the development of an intercultural philosophy. Globalization has to face the problem of the universe and its relation to the particular cultures. Some philosophers argue that *universality* is sharply opposed to *particularity* (i.e., there is a "*break*" in culture). Others allow for both the universal and particular focusing on their interrelation, arguing that diversity and multiculturality do not exclude forms of cultural unity. The investigation involves the current debate regarding the term "*intercultural philosophy*" itself. For some philosophers, the term "intercultural" seems to be incompatible with philosophy as universal knowledge.

Well known European philosophers concerned with intercultural philosophy are: *Rauél Fornet—Betancourt* (b.1946) who has studied Hispanic, African and Latin-American cultures, *Ram Adhar Mall* (b.1937) who has studied Indian philosophy, *Frauz Martin Wimmer* (b.1942) who thinks that philosophy should be rewritten to include other than European traditions, and has developed ways for intercultural dialogues, called "*polylogues*" (many fold dialogues) by him, and *Heinz Kimmerle* (b.1930) who has departed from colonial thinking toward a dialogue with African philosophy based on complete equity.

The *polylogue* approach to intercultural philosophy includes methods that disable any unreasonable universalism or relativistic particularism. The fundamental rule in this approach is: "never consider the philosophical arguments of an author of a specific cultural tradition as well founded", and the governing principle of intercultural hermeneutics is the classical principle of "*equity*" [12]. In intercultural philosophy, dogmatic ideas and the supposition that ethical and ethnic differences are correlated, are not acceptable. Elmar Hollenstein has proposed a set of rules of thumb that help to avoid misunderstandings in intercultural dialogues [13]. Actually, the concept of international dialogue/polylogue is considered as a "*regulative idea*" in creating an alternative to current globalization [14]. Questions of philosophy, questions about the fundamental structures of reality, the knowledge-structures of reality, the knowledge-ability, and the validity of norms have to be discussed and addressed in such a way that an answer/solution is not propagated unless a polylogue among as many traditions as possible has taken place. This recognizes the relativity of concepts and methods, and it implies a *non-centralized* view to the history of human thinking. Clearly, to grasp the differences among philosophical views of several cultures one has to look at the whole of all views and not to what is common to all of them, because then the result obtained may be "void". In comparative philosophy the dialogue is intercultural, and finally transcultural (not simply an inner dialogue), that goes beyond any noncultural foundation of philosophy, while at the same time remaining attached to it for the articulation of the different opinions. According to Heidegger [15], when "*we*" get into dialogue in the European tradition (as originated by the Greek experience) the meaning of the word "*we*" should be enlarged to take an intercultural interpretation.

Two reputable journals on intercultural philosophizing are: *Polylog* (published in Vienna, Austria), and *Simplegadi* (published in Padua, Italy). The journal Polylog (http://prof.poly-og.org) defines intercultural philosophy and its prospects as follows:

> We understand intercultural philosophy as the endeavor to give expression to the many voices of philosophy in their respective cultural contexts and thereby to generate a shared, fruitful discussion granting equal rights to all. In intercultural philosophy we see above all a new orientation and new practice of philosophy – of a philosophy that requires an attitude of mutual respect, listening, and learning. It entails a new orientation because, in acknowledgement of the cultural situatedness of philosophy, claims must prove themselves interculturally, and culture and cultures must be consciously kept in view as the context of philosophizing. It entails a new practice because this consciousness demands a departure from an individual, mono-cultural production of philosophy and seeks instead a dialogical, process-oriented, fundamentally open polyphony of cultures and disciplines.

A more general journal which is concerned with philosophical, social, moral, and political issues of multiculturalism is the journal *Diversities,* published by UNESCO-*Social and Human Sciences*. A forum for intercultural philosophy, which includes an anthology, themes, archive, and literature, is available at www.polylog.org. A list of free online journals on intercultural philosophy, and the print editions of the Austrian journal Polylog and the Indian Journal for Intercultural Philosophy is available at www.link.polylog.org/jour-en.htm.

10.5 Intercultural Issues of Infoethics and Roboethics

Ethical concerns in relation to *information and computer technology* (ICT) and robotics are becoming important global/intercultutal issues. These issues can be addressed easily by applying intercultural philosophy concepts and principles. Ideally we would like to have universal principles to handle the ethical problems arising in robotics and ICT in global intercultural contexts. The question here is "whether this is possible". Many scholars have argued that IIE (*Intercultural infoethics*) and IRE (*intercultural roboethics*) are dominated by the Western philosophical ideas and practices which may not immediately be compatible with Far-East and other traditions. For example, the arguments for the protection of privacy are based on the Western concept of autonomy (*autonomy of the individual*) which is different than the Confucian-based (Japanese) concept of "*collective common good*" over and above the benefit of individuals.

According to Ess [16, 17] the aim of IIE/IRE is as follows:

- To study both local and global IIE/IRE aspects such that to respect local values, traditions, preferences, etc.
- To provide shared universal or almost universal solutions to control ethical problems arised from robotics and information technologies.

As Wong points out [18], the above aim is ambiguous because:

- It is not clear what the meaning of "respect local values and traditions" is.
- It is not clear what "shared universal or almost universal solutions" means.

Two possible meanings of the aim of IIE/IRE as characterized by Ess are:

- Advocate "shared norms-different interpretations".
- Advocate "shared norms-different justifications".

Wong argues that the first meaning is untenable, and the second is acceptable only with qualifications [18]. To overcome the inadequacy, he suggests an alternative definition of the aim of IIE/IRE, namely "*to establish a set of shared values*", instead of establishing shared norms.

According to Himma [19] two distinct stages can and must be followed for defining an intercultural ethics framework, namely:

- Descriptive analysis of different moral systems in various cultures (empirical findings)
- Normative analysis of these moral systems and the corresponding goal to formulate universal (or quasi-universal) moral principles for facing ICT/robotics-related ethical issues.

Descriptive analysis includes tasks such as explicating moral norms/moral values embodied in various cultural traditions, and analyzing the impacts of ICT/robotics to these cultures. The empirical findings will then provide the foundation for determining the universal (or quasi-universal) moral principles.

Normative analysis provides normative and evaluative judgments for formulating the ethical issues derived from a specific cultural perspective, and at the same time provides shared universal solutions to control ethical problems.

Clearly, both descriptive (empirical) and normative analyses are necessary to be carried-out in order to achieve a sufficient IIE/IRE system for criticizing and sentence those who do not comply with the standards as specified.

As explained in [18] the justifications in the "shared norms, different justifications" framework must not be pragmatic, because the use of pragmatic justifications runs against IIE/IRE. On the other hand, moral justifications are based on moral values within a particular moral framework. Therefore, neither form of justifications help the "shared norms-different justifications" approach to satisfy the fundamental requirements of IIE/IRE. The concern about ethical justifications springs from the complexity of various cultural traditions.

In Western ethics a norm is typically justified in a utilitarian way, whereas the negation of the same norm is justified by deontological arguments. Similarly, in Eastern (Confucian) ethics the "*canons*" do not lead to fixed rules, but to a school of thought that involves various sub-traditions (neo-Confucianism, Daoist, Zen) which have their own moral systems. The problem here is that when different ethical justifications are equally legitimate (i.e., equally justifying the norms) there is the possibility that no norm will ever be shared. This means that a hierarchical order of the ethical justifications is required for the "shared norms, different justifications" approach to work. Otherwise the prospect of shared norms is fuzzy and does not suggest uniquely which norm should be shared.

To face this ambiguity, Wong [18] proposes to use a "*set of shared values*", i.e., to follow a value-based approach instead of a norm-based approach. Of course this implies, that we need to identify *common values* which are valid across cultures and guarantee human (and non human) flourishing. Actually, this set of fundamental common values must be defined *normatively* and maintained/promoted as far as possible. The problem in this normative approach for determining ethical aspects based on the "shared values", is how these values are mapped to ICT/robotics-related issues in IIE/IRE. Since these ethical issues arise from completely different cultures, with very different values, a deep examination of several scenarios and the values contained therein is needed as suggested by the Polylogue theory. It may be that no shared values exist. However, in any case looking at the values helps to face issues that are marginal in the norms-based approach, such as gender, well-being, and digital divide issues.

Returning to our discussion of Western and Japanese roboethics we recall that there are invisible reasons which lie behind the differences in these two traditions. Western roboethics is based on the Western understanding of the concepts of autonomy and responsibility. Japanese have difficulty to understand "autonomy" and "responsibility" of robots. As already pointed out, this is due to that they have a different kind of shared and normalized frame of narratives, stories and novels. Japanese people develop strong emotional sensitivity to persons, things and events in life, which influences their little interest in "abstract discussions" and direct emotional expressions with regard to robots and ICTs. In Japan, robots seem to be created with some kind of images such as [1]:

- Iyashi (healing, calmness)
- Kawai (cute)
- Hukushimu (living)
- Nagyaka (harmonious, gentle)
- Kizutuku-Kokoro (sensitive inner minds)

These images cannot be separated from Japanese "*intersubjective sensitivity*" or "*emotional place*". In other words, Japanese robots appear to be interacting with people in cultural contexts where abstract concepts and discussions have very much less importance than communication and interaction based on indirectly originated feelings and emotions. Nagata [1] notes that in other Eastern (Asian) cultures (China, Korea, etc.) more direct/straightforward emotional expressions are considered to be better, compared to Japanese people who are accustomed to indirect emotional expressions. These differences often cause many misunderstandings among people in different Eastern countries, which tend to have a negative image about Japanese culture. These people feel that although Japanese people look gentle, actually they are not friendly, and that there are invisible barriers between them and Japanese people, due to different cultural contexts [1]. Some further issues of intercultural infoethics, including a comparative analysis of P2P software usage in Japan and Sweden, are provided in [20].

10.6 Robot Legislation

Here, a short discussion of the robot related legislation in Europe and South Korea will be provided, along with some comments on the robotics legislation in Japan, China and USA.

In Europe (and the West) civil liability in the manufacture and use of robots is distinguished in [21]:

- Contractual liability
- Non-contractual liability

Contractual liability regards the robot as the object (product) of the contract for sale between the manufacturer (seller) and the user (buyer), i.e., the robot is considered to be a "consumer good, product or commodity". Here, the standard liability rules are applicable without any difficulty, i.e., in the West the existing legislation seems to cover the case where the objects are robots without requiring any addition or modification. Contractual liability occurs when the robot performance is not as agreed in the contract, even when the robot does not make any damage or harm. Two documents about the European legislation referring to commercial objects are: PECL (Principles of European Contract Law) and CECL (Commission on European Contract Law).

Non-contractual liability occurs when the robot's action causes a "legally recognized damage" (e.g., infringement of a human right) to a human regardless of the existence of a contract.

Two cases may occur:

- *Damage caused by a defective robot*:
  In this case we have the so-called "objective liability" of the manufacturer due to a defective product/robot (i.e., liability without fault). If more manufacturers or suppliers and distributors are involved, the liability is jointly assigned to all of them (European Directive 85/374).
- *Damage caused by action or reaction of the robot with humans:*
  Frequently, this may be due to the learning mechanism embedded in the robot, which involves some kind of unpredictable behavior. Situations like this have been treated in USA in analogy to the case of an animal and a moving object.

Currently, robots (machines) in the West do not have the legal status of moral agent (like the human). In future, an autonomous intelligent robot may be regarded as a *legal person* in analogy to companies or corporations, and so they may enter a public register (similar to the commercial register).

In South Korea a special law on the development and distribution of intelligent robots has been launched which is applied in association with the *Korean Robot Ethics Charta* (2005). The robot is defined as "a mechanical device that perceives the external environments for its self, discerns circumstances, and moves voluntarily". This law involves two sections, namely [22]:

- Quality certification of intelligent robots section.
- Insurance section.

In the first section, the *Minister of Knowledge Economy* (MKE) authorizes a "*certifying institution*" to issue "certificates" of the quality of intelligent robots, formulates the policy for distribution and dissemination of certified robots, and provides the legislation concerning the designation, cancellation of designation, and operation of a certifying institution.

The second section defines the persons that may operate a business for the purposes of insuring damage produced on consumers by certified intelligent robots. All the above provisions are prescribed by Presidential Degrees.

The *Korean Robot Ethics Charta*, issued and adopted by the respective Charter, provides a set of principles about the human/robot ethics that assure the co-prosperous symbiosis of humans and robots. These principles are the following [22]:

- *Common human-robot ethics principle* (The human being and the robot are both deserved of dignity, information, and engineering ethics of life).
- *Robot ethics principle* (The robot should obey the human as a friend, helper, and partner and should not harm human beings).
- *Manufacturer ethics principle* (Robots should be, manufactured such that to defend human being dignity. Manufacturers are responsible for robot recycling and providing information about robot protection).
- *User ethics principle* (Robot users must regard robots as their friends and forbid any illegal re-assembly or misappropriation of a robot).
- *Government ethics principle* (The government and local authorities must enforce the effective management of robot ethics over all the manufacturing and usage cycle).

One can easily observe that the above Korean human-robot ethical principles have an emotional and social character rather than a legal one. The author is not aware of the existence of analogous solid human-robot ethics Charta in Western countries. In the West, ethical issues of robots society are resolved by combinations of deontological, utilitarian, and casuistry theories, along with professional ethics codes (NSPE, IEEE, ASME, AMA, etc.).

From the above it is evident that the development of a "robotics law and ethics Charta" is urgently required in Europe (and the West). A noticeable effort towards this end is the recently launched (March 2012) EU-FP7 *Project ROBOLAW* (Regulating Emerging Robotic Technologies in Europe—Robotics Facing Law and Ethics). This project aims at providing the European Union with a "*White Book on Regulating Robotics*", the ultimate goal of which is the establishment, in the near future, of a general solid framework of robot law in Europe. The project is coordinated by the "Scuola Superiori Sant'Anna" (Italy) with partner Institutions across Europe specialized in robotics, ethics, assistive technology, law, and philosophy.

This project is exploring the interrelations among technological, legal and moral issues in the field, in order to promote a legally and ethically sound basis for the robotics achievements of the future (www.robolaw.eu). The project is supported by a "Network of Stakeholders" (Disabled People Associations, Care Givers Associations, Producers of assistive and medical robots, Standardization Bodies (ISO), Trade, Unions, Insurance Companies, etc.).

Legislation on surgical robots (and, more generally on medical devices) in European Union is being reviewed, since in most Member States it is currently incomplete. This field is closely related and partly covered by the EU Directive on the Application of patients' rights in cross-border health care. An important part of EU legislation update is the Creation of a *Central European Data Bank* (like that of Japan) to enable reporting all harmful incidents and reclassifying certain products in Class III. To this end, all *Unique Device Identification* (UDI) code which will allow an effective traceability of medical devices.

As mentioned in Sect. 5.7, in Japan a central database system already exists for logging and communicating any injuries caused to humans by robots. Japan is particularly concerned about the human safety in using robots, and has already the relevant legislation.

In China, there is still a lack of interest in robot-related legislation which obviously is a problem that has to be urgently addressed. With 1.4 billion population, many Chinese have the opinion that their country does not need service and humanoid robots to replace humans. However, the widespread use of autonomous lethal robots by USA, forces China to consider seriously the legal implications of their use.

In USA there are already some robot-specific laws and regulations. For example, In Texas a bill outlaw has been enacted that says: "A person commits an offense if the person uses or authorizes the use of an unmanned vehicle or aircraft to capture an image without the express consent of the person who owns or lawfully occupies the real property captured in the image" (http://robots.net/article/3542.htm). Thus, any robot (in the air, underwater, on the ground), even if operates on public property that inadvertently records any kind of sensor data (sound, visible, thermal, infrared, ultraviolet) originating on private property is deemed illegal. This bill was strongly criticized as "overly broad" and "bad worded", since it may outlaw most outdoor hobby robotic activities and even stop university programs, but seems to exempt federal, state and local police spying under various circumstances. Very recently, California has enacted a law that would substantially restrict places where it is legal to fly recreational drones. "Piloting one inside the boundaries of a property owner's airspace would be considered trespassing" (IEEE Spectrum TechAlert, 12 February 2015).

10.7 Further Issues and Concluding Remarks

Japan is one of the leading countries developing and using robots. It is actually the leader in the development of humanoid and socialized robots. Sociologically and culturally Japan is considered to be the "other" for the Western societies. In Japan a robot is a machine from the substantial point of view. In the West, the key theories of ethics are deontology theory and utilitarian theory. In the East, roboethics considers the robot as one more partner in the global interaction of humans and things (*Mono, Koto, Kokoro*). The moral traditions in Japan are *Seken/Giri* (traditional Japanese morality), and *Ikai* (old animistic tradition). Robots in the Japanese society are "machines", and, when these machines are used by humans, an emotional bound and harmonic relation is created between the robot and the human user. In Japan there is a *desire for relation* since it is believed that a human being is an element of nature like a stone, a mountain or an animal, in which the artifacts are also integrated on an *equity* basis. In the West there is a desire of *hierarchical order* of existence, i.e., humans are on the top of the hierarchy, artifacts at the bottom, and animals between these two.

The production of robots in Japan is motivated and included in the aesthetic struggle for beauty which is characteristic of the Japanese spirit. The Japanese spirit is never threatened by robots as long as robots adapt to technology successfully. The main problem of traditional disregard of the *world order* in Japan, is that a safeguard against a cyborg and/or mechanization of human reasoning is not offered. In Japan the meaning of autonomy is considerably different from that of the West, and the Western concept of "freewill" is hardly found. In other words, the harmonious symbiosis with other is respected, and autonomy can be easily disregarded. People must behave like the other persons in the Society.

In an effort to establish a basis for equity between humans and robots, and confidence of the future developments of robot technology and the numerous contributions that humanoid robots will make to human, the following "*World Robot Declaration*" was issued in Fukuoka Japan Fair (February 25, 2004). It includes three specific expectations from future robots, and declares five resolutions on what must be done to guarantee the existence of the next-generation robot (www.robotfair2004.com/english/outline.htm).

(A) Expectations for next-generation robots

- *Next-generation robots will be partners that coexist with human beings.*
- *Next-generation robots will assist human beings both physically and psychologically.*
- *Next-generation robots will contribute to the realization of a safe and peaceful society.*

(B) Toward the creation of new markets through next-generation robot technology

- *Resolution of technical issues through the effective use of Special Zones for Robot Development and Test.*

- *Promotion of public acceptability of robots through the establishment of standards and upgrading of the environment.*
- *Stimulation of adoption through promotion of introduction of robots by public organizations.*
- *Dissemination of new technologies related to robots.*
- *Promotion of the development of robot technology by small enterprises, and their entry into the robot business. The government and academia shall provide active support for such efforts.*

In the content study presented in [1] it was verified that Japanese people of today still live in the traditional aspect of life world "*Seken*", which is based on *Buddhism, Shintoism, Conficianism* and historical memories. In this study important terms like *destiny,* and *sincerity* are seen to appear with very small frequencies. **Shintoism** has been described in Sect. 10.2.1. **Buddhism** is a religion indigenous to the Indian subcontinent, based on the teachings of *Siddharta Gautama,* known as *Buddha.* The two main branches of Buddishm are:

- *Theravada*: The School of Elders.
- *Mayahana*: The Great Vehicle.

The foundations of Buddhist tradition and practice are the "*Three Jewels*": the *Budha*, the *Dharma* (the teachings) and the *Sangha* (the Society). Refuging in the "Three Jewels" is a declaration and commitment to being a Buddhist, which distinguishes a Buddhist from a non-Buddhist. Ethical practices of Buddhist include: support of the monastic community, and cultivation of higher wisdom and discernment.

Confucianism is a philosophical and ethical system oriented to practical issues, particularly to the importance of family, and is based on the belief that humans are teachable, improvable and perfectible through personal self-cultivation and self-reaction. Confucian ethical concepts and practices include:

- *Ren*: An obligation to altruism and good.
- *Yi*: The upholding of rightness and moral disposition to do good.
- *Li*: A system of norms and propriety that determines how a person should properly act in everyday life.

The cardinal moral values are '*ren*' and '*yi*' and any one that fails to respect them is socially contemptible. Neo-Confucianism, which is oriented towards philosopher Herbert Fingarette is called the "*secular as sacred*".

The culture in the East is not uniform. Humanoid robots are accepted in Japan, China and Korea but not in nations with Islamic tradition (e.g., Indonesia). Buddhist culture challenges the Western idea of a "*coherent self*". In the West the search of universal values is still ongoing, and infoethics/roboethics are considered as observer-dependent reflection on moral norms. From the above, it follows that a continuous dialogue between East and West is needed.

Japan's isolation was ended by the *Meiji Restoration* in 1968 which marked the downfull of the *Tokugawa* regime. The two political slogans during the transition from *Tokugawa* to *Meji* were *bunnei-Kaika* (westernization) and *fukoku-kyohei* (strong military, rich state). Fukuoku-kyohei military policy has succeeded to make Japan a powerful country, whereas by bunnei-kaika policy many Western ideas were transferred to Japan and new terms were invented such as [2]:

- Shakai (society)
- Tetsugaku (philosophy)
- Risei (reasoning)
- Kagaku (science)
- Gijyutsu (technology)
- Shizen (nature)
- Rinri (ethics).

References

1. Nakada M (2010) Different discussions on roboethics and information ethics based on different cultural contexts (Ba), In: Sudweeks F, Hrachovec H, Ess C (eds) Proceeding conference on cultural attitudes towards communication and technology, Murdoch University, Australia, pp 300–314
2. Kitano N (2007) Animism, Rinri, modernization: the base of japanese robotics, workshop on roboethics: ICRA'07, Rome, 10–14 Apr 2007
3. Nakada M, Tamura T (2005) Japanese conceptions of privacy: an intercultural perspective, ethics and information technology, 1 Mar 2005
4. Yoshida M Giri: a Japanese indigenous concept (as edited by L.A. Makela), http://academic.csuohio.edu/makelaa/history/courses/his373/giri.html
5. Kitano N (2005) Roboethics: a comparative analysis of social acceptance of robots between the West and Japan, Waseda J Soc Sci 6
6. Krebs S (2008) On the anticipation of ethical conflicts between humans and robots in Japanese mangas. Int Rev Info Ethics 6(12):63–68
7. South Korea to create code of ethics for robots, (www.canada.com/edmontonjournal/news/story.html?id=a31f6)
8. Lovgren S (2010) Robot code of ethics to prevent android abuse, protect humans. (http://news.nationalgeographic.com/news/2007/03/070316-robot-ethics_2.html)
9. Galvan JM (2003) On Technoethics. IEEE Robotics and Automation Magazine 6(4):58–63, www.eticaepolitica.net/tecnoetica/jmg_technoethics[en].pdf
10. Capurro R (2008) Intercultural information ethics. In: Himma KE, Herman T (eds) Handbook of information and computer ethics. Wiley, Hoboken, pp 639–665
11. Mall RA (2000) Intercultural philosophy. Rowman and Little field Publishers, Boston
12. Wimmer FM (1996) Is intercultural philosophy a new branch or a new orientation in philosophy? In: Farnet-Betacourt R (ed) Kulturen der Philosophie. Augustinus, Aachen, pp 101–118
13. Hollenstein E A dozen rules of thumb for avoiding intercultural misunderstandings, Polylog, http://them.polylog.org/4/ahe-en.htm
14. Demenchonok E (2003) Intercultural philosophy. In: Proceedings of 21st world congress of philosophy, vol 7. Istanbul, Turkey pp 27–31
15. Heidegger M (2008) Intercultural information ethics. In: Himma KE, Herman T (eds) Handbook of information and computer ethics. Wiley, Hoboken

16. Ess C (2007) Cybernetic pluralism in an emerging global information and computer ethics. Int Rev Info Ethics 7:94–123
17. Ess C (2008) Culture and global network : hope for a global ethics? In: van den Haven J, Weckert J (eds) Information technology and moral philosophy. Cambridge University Press, Cambridge, pp 195–225
18. Wong P-H (2009) What should we share? Understanding the aim of intercultural information ethics. In: Proceedings of AP-CAP, Tokyo-Japan, 1–2 Oct 2009
19. Himma KE (2008) The intercultural ethics agenda from the point of view of a more objectivist. J. Inf Commun Ethics Soc 6(2):101–115
20. Hongladarom S, Britz J (eds) Intercultural information ethics. Special issue: international review of information ethics, vol 13, no 10
21. Leroux C (2012) EU robotics coordination action: a green paper on legal issues in robotics. In: Proceeding of international workshop on autonomics and legal implications, Berlin, 2 Nov 2012
22. Hilgendorf E, Kim M (2012) Legal regulation of autonomous systems in South Korea on the example of robot legislation. In: International workshop on autonomics and legal implications', Berlin, Germany, 2 Nov 2012, (http://gccsr.org/node/685)

Chapter 11
Additional Roboethics Issues

> *The danger of the past was that men became slaves. The danger of the future is that man may become robots.*
>
> Erick Fromm

> *We're seeing the arrival of conversational robots that can walk in our world. It's a golden age of invention.*
>
> Donald Norman

11.1 Introduction

In previous chapters we have discussed the ethical aspects of currently widely used robot categories, namely medical robots, assistive robots, social robots, and war robots. Medical, assistive and social robots are bounded by a core of similar ethnical principles referred to autonomy, beneficence, non-maleficence, justice, truthfulness, and dignity. War robots are subject to the international law of war (jus ad belum, jus in bello, jus post bellum). During the last two decades robots have undergone a 'demographic explosion' with the number of service (medical, assistive, home, social) robots exhibiting almost one order of magnitude higher growth than industrial *robots*[†1] as predicted by Engelberger, the father of robotics. Today the robot expansion came to a level in which robots are no longer 'slave machines' that satisfy only human desires, but embody some degree of autonomy, intelligence, and conscience, approaching the so-called '*mental machine*'. The key ethical and legal issue in mental robots is the 'responsibility' aspect, i.e., the problem of assigning responsibility to the manufacturer, designer, user of the robot or to the robot itself in case of harm. As we have already seen, it is almost generally accepted that (so far) robots cannot themselves be held morally responsible, because they

---

[1]IFR Statistical Department Executive Summary of World Robotics: Industrial Robot and Service Robots report (www.worldrobotics.org).

S.G. Tzafestas, *Roboethics*, Intelligent Systems, Control and Automation: Science and Engineering 79, DOI 10.1007/978-3-319-21714-7_11

lack intentionality. Robots can only be regarded as an additional entity in ascribing moral and legal responsibility (see Sect. 6.4).

The purpose of this chapter is to discuss three more important domains of roboethics, namely:

- Autonomous cars
- Cyborgs
- Privacy

11.2 Autonomous Cars Issues

Autonomous (*self-driving, driverless*) cars are on the way. *Google's* driverless cars are already street legal in California, Florida and Nevada (Fig. 11.1). The advocators of driverless cars argue that within two or three decades these autonomously driving cars will be so accurate and safe that will dominate in number human-driving cars [1, 2]. At the basic level, autonomous cars use a set of cameras, lasers and sensors located around the vehicle for detecting obstacles, and through GPS (global positioning systems) help them to move at a preset route. These

Fig. 11.1 Google self-driving cars. *Source* (**a**) http://cdn.michiganautolaw.com/wp-content/uploads/2014/07/Google-driverless-car.jpg, (**b**) http://www.engineering.com/portals/0/BlogFiles/GoogCar.png

devices and systems give the car an accurate picture of its environment, so it can see the layout of the road ahead, if a pedestrian steps out or if the car in front slows or comes to a halt. Roboticists and car manufacturers are trying to develop autonomous cars that will be smother and safer than cars driven by expert and professional drivers. Less time will be spent in near-collision situations, and self-driving cars are expected to accelerate and brake significantly more sharply that they do when they are human-driven.

In U.S.A., Florida was the first state that permitted experimental driverless vehicles to be travelling on public roads, testing how their crash-averting sensors react to sudden and vicious thunderstorms.

It is noted that automated driving cars (at least to a certain degree) already exist today, e.g., cars with collision avoidance systems, emergency braking systems, and lane-change warnings. Ideally, a driverless car is a robot vehicle with nobody behind the wheel that functions accurately and safely on all roads under all conditions.

Scientists and engineers are now not debating whether self-driven cars will come to be, but how it will happen and what it will mean, i.e., discussions are dominated not about technology itself, but about the behavioral, legal, ethical, economic, environmental, and policy implications. In principle, facing the above and other societal issues seems to be more delicate and difficult, than the fundamental engineering problem of driverless car design. Here, the fundamental ethical and liability question is: 'Who will be liable when a driverless car crashes?'

This question is analogous to the ethical/liability question of robotic surgery discussed in Sect. 6.4. The great majority of car collisions today are the fault of one driver or the other, or the two in some shared responsibility. Only few collisions are deemed the responsibility of the car itself or its manufacturer. However, this will not be the same if a car drives itself. Actually, it will be a lot harder to conventionally blame one driver or the other. The ethical and legal responsibility should be shared by the manufacturer or multiple manufacturers, and the people who made the hardware or software? Or the mapping platform, or blame another car that sent a faulty signal on the highway?

Nevada and California enacted legislation permitting self-driving cars to be on the roads. In both legislations it is required that a human be present in the car, sitting in the driver's sit, and able to take the control of the car at any time. Analogous, highway legislation was enacted in Britain [3]. Although no accident has occurred during the testing of Google's driverless cars in Nevada and California, it would be inevitable that a collision will someday occur as their use becomes more popular. Such an occurrence would present very many new issues for the existing framework of responsibility.

Consider the following scenario discussed in [4]:

'Michael is the back driver in a self-driving car. He sits behind the wheel as required by the law, and pays attention to his surrounding. It begins to rain lightly. The car advises that under inclement weather conditions, a driver must manually control the vehicle. Because the rain is not heavy, Michael believes it does not raise to the level of being 'inclement weather', so he allows the car to continue driving without assistance. The car suddenly makes a sharp turn and crashes into a tree, injuring Michael'.

The question that arises here is: 'who is in fault?' Should Michael have taken the driving wheel when it started raining, or was the car's instruction too fuzzy to impose that responsibility to him? Michael would likely sue the vehicle's manufacturer under the theory of 'products liability'. The manufacturer would argue that Michael had a duty to control the car manually when it began to rain. In this scenario, only Michael himself was injured, but how would responsibility be distributed if a third party had been injured as a result? Actually, here there is no clearly applicable ethical rule to determine how Michael should have acted, and also the available legal framework is less clearly defined.

Other questions related to the use of autonomously driven cars are the following [5]:

- How will people still driving an old-fashion car behave around autonomous vehicles when there is still a mix of the two on the road? Some human drivers may behave more venturesome around self driven cars—weaving or speeding around them—because they expect autonomous cars to correct their behavior.
- What if your autonomous car does not drive like you do? For example if you are one of those people who drives excessively slowly on the highway, how will you react when you are sitting in the front seat of a car that drives faster than you are used to do?
- Vice versa. Will people be fascinated to take control from these vehicles? How can we learn to trust automated vehicles?
- If for some reason a vehicle requires you to suddenly take wheel, will you be able to quickly turn your attention away from what you were doing while your car was doing the driving for you?
- Will consumers want to buy driverless cars, appreciating the benefits involved (e.g., safer driving, congestion reduction, cut down on the amount of source urban land needed for parking, etc.)?
- How can we find a way to align the costs and benefits of driverless cars for the people who might buy them?
- How will autonomous cars change our travel and consumption pattern? For example, if these cars make travel easier, perhaps they will induce new trips that we are not making today, thus increasing the number of trips and kilometers we collectively travel.
- Is technological infrastructure ready? Particular questions here are: (i) What kind of lighting is needed on city streets if we are trying to optimize for radar vision instead of human sight? (ii) Can a computer process a street sign which is covered by graffiti? (iii) Will car manufacturers want to make autonomous vehicles if only a few places of a country are ready for them?

A dominant issue in adopting widely self-driving cars is related to the communication link. Vehicle-to-vehicle driverless communication, which lets car tell each other what they are doing so they won't collide, may be headed for political difficulties [6]. It requires a big portion of broadband, which the Federal Communications

Commission (FCC) set aside for carmakers in 1991. But the expansion of smart phones and video devices that stream movies and other video has absorbed much of the broadband spectrum. Now, the big cable companies have banded together to lobby congress to let them share in the part reserved for automobiles. The danger that could be created if the devices used by cable companies cause interference with cars could lead to disaster. A dropped call on a cell phone caused by interference is not a big deal, but the loss of even a little data on a vehicle's collision-avoidance system could be fatal. A discussion of the consequences of self-driving cars is presented in [7]. Some scientists, engineers and thinkers argue that driverless cars should be slow down. For example, *Bryan Reimer,* a researcher at MIT says that one fatal crash involving a self-driving vehicle would become front page news, 'shut down the robotic industry'-and lead automakers to a 'major pullback in automatic safety systems' like collision avoidance technology going in conventional cars now. Professor *Raja Parasuramam* (George Mason University) says that 'there will always be a set of circumstances that was not expected, that the automation either was not designed to handle or other things that just cannot be predicted'. However, despite widespread concern, no carmaker wants to be left behind when it comes to at least partially autonomous cars. They believe that this is going to be a technology that will change humanity. It is a revolutionary technology, and despite some people call it disruptive, it will change the world, save lives, save time, and save money. Another, more extensive discussion on the positive and negative implications of self-driving vehicles is provided in [8].

11.3 Cyborg Technology Issues

Cyborg technology is concerned with the design and study of neuromotor prostheses aiming at restoring and reinstating lost function with replacement that is different as little as possible from the real thing (a lost hand or arm, lost vision, etc.) [9–11]. Neuromotor prostheses allow disabled people to move purely through the power of the brain and in the long-term their recipients will be able to '*feel*' through.

The word *cyborg* stands for *cybernetic organism,* and more broadly refers to the concept of *bionic man*. The cyborg technology has been made possible by the fact that the brain central nervous system bioelectrical signals can be connected directly to computers and robot parts that are either outside the body or implanted in the body. Clearly, making an artificial hand with the movement and sensors needed to creating an entity that approaches the capability for movement and feeling of a normal functioning biological hand is an extremely advanced attainment. Of course, when one succeeds in decoding the brain's movement signals, then these signals can be directly connected to external electronic equipment (mobile phones, TV sets, etc), such that it may be possible to control electronic devices with the power of thought alone i.e., without the use of articulated language or external motor device. In other words, cyborg-type prostheses can also be *virtual.* In more broad terms, the concept '*cybernetic organism*' is used to describe larger communication and control networks, e.g., networks of roads, networks of software, corporations, governments, cities, and the mixture of these things.

In medicine, there are two principal types of cyborgs, namely the *restorative* and the *enhanced*. Restoring technologies 'restore lost function, organs, and limbs' such that to revert to a healthy or average level of function. The enhanced cyborg technologies follow the principle of optimal performance, i.e., maximizing output (the information or modification achieved) and minimizing input (the energy expended in the process).

Cyborg technology is a *two sided sword*; it can be used for the good or for the bad, and like other modern technologies may have negative and positive consequences.

Some negative arguments about cyborg (bioman) technology are the following [9]:

- Human race will divide along the lines of biological haves and have-nots. For example people with sufficient money will be possible to augment their personal capabilities as they think fit (as, e.g., it is already done with spas, plastic surgery, etc.), as well as to utilize organ replacement, etc.
- The military can use this technology to get super-soldiers with faster reflexes, fatal accuracy, less fatigue, etc.
- Many people anticipate tremendous possible risks towards health an safety (such as those currently induced by Prozac, silicone breast implants, steroids, artificial hearts, etc).
- With these enhancements some people may become capable of obtaining incomes and greater opportunity on the labor market than most others.
- More seriously, cyborg technology (bioelectronics) could lead to the ability to monitor and control people.

In society, cyborg technologies could be used in two areas, namely [12]:

- The public health service area which has healing and therapeutic purpose (e.g., to remedy injuries that cause disabilities).
- The finance and private market area where cyborg-technology-based supplements are used for the purpose of enhancement, possibly restricted by related legislation.

In finance, due to the advances of information and control technology, investors have available for use supercomputers to engage in trading, banking, brokering, and money management. Actually, modern finance is becoming '*cyborg finance*' because the key players are partly human and partly machine. A major characteristic of cyborg finance is the use of extremely powerful and fast computers to investigate and execute trading opportunities based on complex and mathematical models. The software developed on the basis of such complex algorithm is in most cases proprietary and non-transparent. For this reason cyborg trading is often referred to as '*black-box trading*'. Cyborg ethics includes the ethics of medicine, medical roboethics, and assistive roboethics (especially prosthetic roboethics). The primary ethical questions surrounding the development of cyborgs are focused on human dignity, human relations, protection of physical/bodily harm, and the management of the health and other personal data evaluation. In all cases the primary six rules of roboethics should be respected (autonomy, non-maleficence, beneficence, justice, truthfulness, dignity).

The *Danish Council of Ethics* has issued two legislation recommendations for the use of cyborg technology in society. These are the following:

A relatively loosely regulated framework According to this framework, every person in the society should be free to decide the ways in which he/she wishes to exploit the opportunities offered by cyborg technologies, as much as this decision does not damage the life development or private lives of others. Of course, the abilities that nature gives to each person imply that individuals have differing conditions for pursuing life goals and desires. An individual with high intelligence (all others being equal) has better opportunities in our society than a person with low intelligence.

A relatively strictly regulated framework Cyborg technology inplanted in or integrated with people should be used exclusively for the purpose of healing disease or remedying disability (e.g., for replacing naturally existing functions—sight, limbs, etc., for people that have lacked them from birth or lost them due to various reasons). Cyborg technology should not give rise of black market. Private hospitals offering cyborg technological interventions and supplements should satisfy the same restrictions and the same valid purposes that apply in the public health system.

The ethical basis for the relatively strict regulatory framework of cyborg technology is that it would have a critically bad effect on fundamental, ethical norms in our society, and that the conditions for a fair relationship between people's life opportunities would be damaged, if it becomes possible for adult persons to purchase cyborg enhancements. Furthermore, cyborg technological enhancements could undermine the human value of '*authenticity*', and could change fundamental characteristics of the human being as species.

The main advantages of mixing organs with mechanical parts are for the human health. Specifically:

- Persons that were subject to surgery for replacing parts of their body (e.g., hip replacement, elbows, knees, wrists, arteries, veins, heart values) can now be classified as cyborgs.
- There are also brain implants based on the neuromorphic model of the brain and the nervous system. For example, there are brain implants that help reverse the most devastating symptoms of the Parkinson Disease. A deaf person can have his inner ear replaced and be able to engage in telephone conversation (or hear music in future).

The disadvantages of cybernetic organisms include the following:

- Robots can sense the world in ways that human cannot (ultraviolet, X-rays, infrared, ultrasonic perception). Thus there is more dependence on cyborg technology.
- Intelligent robots can outperform humans in aspects of memory and mathematical/logic processing.

- Cyborgs do not heal body damage normally, but, instead, body parts are repaired. Replacing broken limbs and damaged armor plating can be expensive and time consuming.
- Cyborgs can think the surrounding world in multiple dimensions, whereas human beings are more restricted in that sense.

A philosophical discussion about cyborgs and the relationship between body and machine is provided in [13], and a general scientific discussion about cyborgs and the future man kind is given in [14].

Some real examples of human cyborgs are the following [15]:

Example 1
The artist *Neil Harbisson* was born with extreme color blindness (achromatopsia) been able to see only black and white. Equipped with an 'eyeborg' (a special electronic eye) which renders perceived colors as sounds on the musical scale he is now capable of experiencing colors beyond the scope of normal human perception. This device allows him to 'hear' color. He started by memorizing the name of each color which was then became a 'perception' (Fig. 11.2).

Example 2
Cyborg technology is useful to replace and ambulated human limb (arm or leg) *because* of illness or injury. *Jesse Sullivan* is a pioneer in this respect, being one of the world's first cyborgs equipped with a bionic limb connected through a nerve-muscle graft. Sullivan is able to control his new limb with his mind, and also to feel hot, cold, and the level of pressure his grip is applying (Fig. 11.3).

Example 3
Jens Naumann has lost his sight (in both eyes) due to a couple of serious accidents. He became the first person in the world to receive an artificial vision system, equipped with an electronic eye connected directly to his visual cortex through brain implants (Fig. 11.4).

Fig. 11.2 Wearing an eyeborg a person with achomatopsia can see colors. *Source* http://www.mnn.com/leaderboard/stories/7-real-life-human-cyborgs

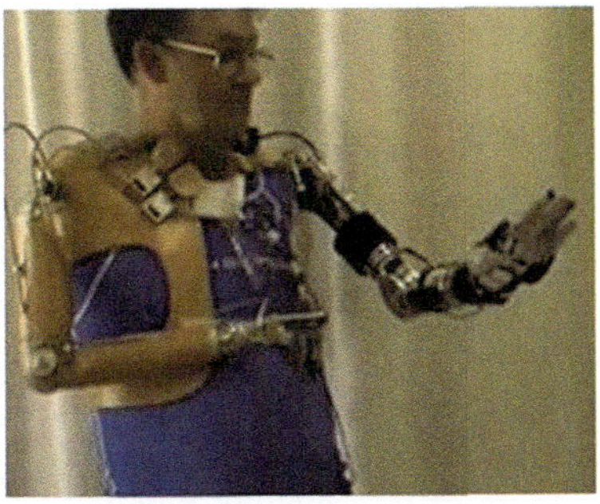

Fig. 11.3 A cyborg limb that can be controlled by a person's mind. *Source* http://www.mnn.com/leaderboard/stories/7-real-life-human-cyborgs

Fig. 11.4 Jens Naumann sees with a cyborg (electronic) eye connected directly to his visual cortex. *Source* http://www.mnn.com/leaderboard/stories/7-real-life-human-cyborgs

Example 4

After losing part of his arm because of an accident at work, *Nigel Ackland* got an upgrade enabling him to control the arm through muscle movement in his remaining forearm (Fig. 11.5).

He can independently move each of his five fingers to grip delicate objects or pour a liquid into a glass. The range of movement achieved is really extraordinary.

Fig. 11.5 A cyborg controlling the arm and fingers through muscle move. *Source* http://www.mnn.com/leaderboard/stories/7-real-life-human-cyborgs

11.4 Privacy Roboethics Issues

Privacy is one of the fundamental human rights. According to the *MacMillan Dictionary,* privacy is defined as the 'freedom to do things without other people watching you or knowing what you are doing'. According to the *Merriam Webster Dictionary*, privacy is 'the state of being away from other people, the state of being away from public attention'. The boundaries and content of what is considered private differ among cultures and individuals, but they have some common aspects. The area of privacy intersects with security domain, which can involve issues of appropriate use and protection of information. Saying that something is private to an *individual,* we usually mean that something is inherently special or sensitive to him/her. The legislation of most countries involves rules and acts aiming to protect the privacy of their citizens, and punishing people that interfere in and observe the privacy of other individuals.

The four types of privacy are the following (Wikipedia):

- *Personal privacy* defined as preventing intrusions into one's physical space of loneliness.
- *Information privacy* or *data privacy* which refers to the changing relationship between technology and the legal and ethical right to privacy in the collection and sharing of data about one's self.
- *Organization privacy* which refers to the desire of governmental agencies, societies or groups, and other organizations to adopt several security practices and control to keep private information confidential.
- *Spiritual and intellectual privacy* which refers to the broader concept of one's property that involves every form of possession (intangible and tangible).

Modern robots which *possess* the ability to sense (via several sophisticated sensors), process, and store the world surrounding them have the potential to implicate human privacy. Robots can move and go to places that humans cannot go, and watch things humans cannot see. All categories of sensor-based robots connected (or not) to computer communications or the Internet (service robots, robots at home, assistance robots, social robots, war robots) may be used for direct surveillance or spying on homes.

Ryan Calo [16] classifies the ways that robots can implicate privacy in the following categories:

- Direct surveillance
- Increased access
- Social meaning

and discusses some ways we can restore and rectify the potential impact of robots on privacy, although, under the current legislation on privacy, this might be very difficult to be done.

To face the increased danger of damaging privacy by robots (equipped with communication and information technologies), our society should enact better and more efficient laws, and technology should develop better engineering security practices.

In particular, social robots implicate privacy in new ways. When an anthropomorphic social robot is used for entertainment, companionship or therapy, the human users develop social bonds with it, and in most cases they do not think that the robot could have a strong effect on their privacy. This is one of the most complex and difficult issues to be faced when using such robots. Several research robotics working in the privacy and security robotic field (e.g., scientists of Oxford University) are trying to find ways for preventing robots from unnecessarily revealing the identities of the people they have captured.

Researchers from the University of Washington have evaluated the security of three consumer-level robots, namely [17, 18]:

- The *WowWee Rovio*, a wireless mobile robot marketed to adults as a home surveillance tool that can be controlled over the Internet and is equipped with a video camera, microphone and speaker (Fig. 11.6).

Fig. 11.6 WowWee Rovio robot. *Source* (**a**) http://www.thegreenhead.com/imgs/wowwee-rovio-robotic-home-sentry-2.jpg, (**b**) http://seattletimes.com/ABPub/2008/01/11/2004120723.jpg, (**c**) http://www.szugizmo.pl/326-thickbox/mini-robosapien-v2.jpg

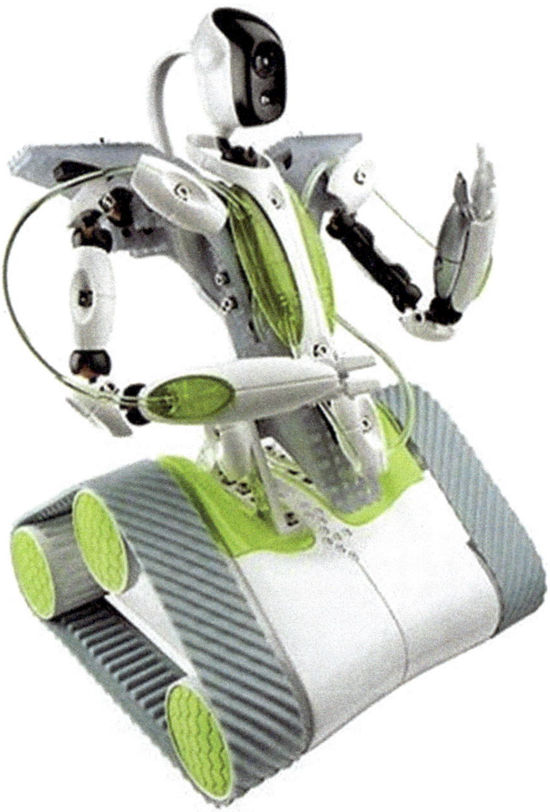

Fig. 11.7 Erector Spykee robot. *Source* http://ecx.images-amazon.com/images/I/31desPZvgQL._SL500_AA300.jpg

- The *Erector Spykee*, a toy wireless Web-controlled 'spy' robot that has a video camera, microphone and speaker (Fig. 11.7).
- The *WowWee Robo Sapien V2*, a dexterous anthropomorphic robot controlled over short distances using an infrared remote control (Fig. 11.8).

One of the worrying security issues these scientists discovered was that the presence of the robot can be easily detected by distinctive messages sent over the home's wireless network, and that the robot's video and audio streams can be intercepted on the home's wireless network, or, in some cases, capsured over the Internet. This weakness may be exploded by a malicious person who could even gain control of the robots (because the usernames and passwords used to access and control the robots are not encrypted except in the case of the Spykee, which only encrypts them when sent over the Internet).

Experts all over the World express their worries (ethical and legal) about the issue that giants like Google, Apple and Amazon are investing in robotics. Recently, Google has acquired many robotics companies. This is exciting for robotics, but the question is 'what is the giant planning to do with this technology?'

Fig. 11.8 WowWee RoboSapienV2. *Source* http://www.techgalerie.de/images/6055/6055_320.jpg

Arner Levin of Ryerson University (Toronto, Canada) posed the question: 'Is there something we should be worried about? If there is, what can we do about it?' [19].

In summary, the key issues of the implication of robots to privacy are the following [16]:

- Robots entering into traditionally protected places (like the home) give the government, private litigants, and hackers increased and wider possibility of access to private places.
- As robots become more socialized (with human-like interaction features), they might reduce people's loneliness, and increase the range of information and sensitive personal data that can be gathered from individuals.
- Using robots, individuals, companies, military and government possess new tools to access information about people, for security, commercial, marketing and other purposes.

- As robot's presence is increasing it may cause delicate privacy harms (including psychological damaging effects which cannot be easily identified, measured and resist).

Robin Murphy and *David D. Woods*, proposed three new robotics laws (beyond Asimov's laws) named: '*The Three Laws of Responsible Robots*' [20]. Their aim was to cover the role of responsibility and autonomy when designing any system in which any single robotic platform operates.

These laws are:

Law 1 A human may not deploy a robot without the human-robot work system meeting the highest legal and professional standards of safety and ethics.
Law 2 A robot must respond to humans as appropriate for their roles.
Law 3 A robot must be endowed with sufficient situated autonomy to protect its own existence as long as such protection provides smooth transfer of control which does not conflict Law 1 and Law 2.

11.5 Concluding Remarks

In this chapter we discussed a number of an additional roboethics issues and concerns that refer to the use of autonomous (driverless) cars and cyborgs, and the issues of privacy security when using home robots and social robots. Another class of robots that rise severe social and ethical questions is the class of *love making robots* (sexbots, fembots, lovobots), for which, as naturally expected, a strong debate is increasingly on going among roboticists, philosophers, sociologists, psychologists, and political scientists. Love robots are social robots that are designed and programmed to mimic and manipulate human emotions in order to evoke loving or amorous reactions from their users. Their design should follow from the early stages the principles of machine ethics and what is known about the philosophy of love, the ethics of loving relationships, and the psychological value of the erotic.

Sullins [21] provides a thorough discussion of the ethical aspects of sexbots concluding by proposing certain ethical limits on the manipulation of human psychology when it comes to building sexbots and in the simulation of love in such machines. He argues that 'the attainment of erotic wisdom is an ethically sound goal and that it provides more to loving relationships than only satisfying physical desire'. Other discussions on the ethical and legal implications of using sexbots are provided in [22–26], which may generate more rounds of discussions on the subject. The opinions expressed range between two extremes. Those in favor say: 'Aye, Robot'. They argue that 'From the perspective of *autonomy,* it is hard to see anything wrong with sex with a robot. According to the *harm* principle, if people want sexbots and they do no harm, there are no grounds, for judging people who use them'. Those that are against sexbots, express several reasons for opposing (if not prohibiting) them. Some of them are the following [26]: "(i) They can promote unhealthy attitudes toward relationships, (ii) They could promote misogyny, (iii) They could be considered prostitutes, (iv) They could encourage gender stereotypes, (v) They may replace real relationships and distance users from their partners, and (vi) Users could develop unhealthy attachments to them". The most extreme opposition to sexbots is expressed in the so-called *Futurama argument*: "If we remove the motivation to find a willing human partner, civilization will collapse. Engaging with a robot sexual partner will remove that motivation. Therefore, if we start having sex with robots, civilization will collapse" [23]. Two examples of sexbots are given in [27, 28].

References

1. Marcus G. Moral machines. http://www.newyorker.com/news_desk/moral_machines
2. Self-driving cars. Absolutely everything you need to know. http://recombu.com/cars/article/self-driving-cars-everything-you-need-to-know
3. Griffin B. Driverless cars on British road in 2015. http://recombu.com/cars/article/driverles-cars-on-british-roads-in-2015
4. Kemp DS Autonomous cars and surgical robots: a discussion of ethical and legal responsibility. http://verdict.justia.com/2012/11/19/autonomous-cars-and-surgical-robots
5. Badger E. Five confounding questions that hold the key to the future of driverless cars. http://www.washigtonpost.com/blogs/wonkblog/wp/2015/01/15/5confoundingquestionsthatholdthekeytothefutureofdriverlesscars
6. Garvin G. Automakers say they'll begin selling cars that can drive themselves by the end of the decade. http://www.miamiherald.com/news/business/article1961480.html
7. O'Donnell J, Mitchell B. USA today. http://www.usatoday.com/story/money/cars/2013/06/10/automakers-develop-self-driving-cars/2391949
8. Notes on autonomous cars: Lesswrong, 24 Jan 2013. http://lesswrong.com/lw/gfv/notes_on_autonomous_cars/
9. Mizrach S. Should be a limit placed on the integration of humans and computers and electronic technology? http://www.2fiu.edu/~mizrachs/cyborg-ethics.html
10. Lynch W (1982) Wilfred implants: reconstructing the human body. Van Nostrand Reinhold, New York, USA
11. Warwick K (2010) Future issues with robots and cyborgs. Stud Ethics Law Technol 4(3):1–18
12. Recommedations concerning Cyborg Technology. http://www.etiskraad.dk/en/Temauuniverser/Homo-Artefakt/Anbefalinger/Udtalelse%20on%20sociale%20roboter.aspx
13. Palese E. Robots and cyborgs: to be or to have a body? Springer Online: 30 May 2012. http://www.nebi.nlm.nih.gov/pmc/articles/PMC3368120/
14. Sai Kumar M (2014) Cyborgs—the future mankind. Int J Sci Eng Res 5(5):414–420. www.ijser.org/onlineResearchPaperViewer.aspx?CYBORGS-THE-FUTURE-MAN-KIND.pdf
15. Seven real-life human cyborgs (Mother Nature Network: MNN), 26 Jan 2015. http://www.mnn.com/leaderboard/stories/7-real-life-human-cyborgs
16. Ryan Calo M (2012) Robots and privacy, in robot ethics: the ethical and social implications of robotics. Lin P, Bekey G, Abney K (eds), MIT Press, Cambridge, MA. http://ssrn.com/abstract=1599189
17. Quick D. Household robots: a burglar's man on the inside. http://www.gizmag.com/household-robot-security-risks/13085/

18. Smith TR, Kohno T (2009) A spotlight on security and privacy risks with future household robots: attacks and lessons. In: Proceedings of 11th international conference on ubiquitous computing. (UbiComp'09), 30 Sept–3 Oct 2009
19. Levin A. Robots podcast: privacy, google, and big deal. http://robohub.org/robots-podcast-privacy-google-and-big-deals/
20. Murphy RR, Woods DD (2009) Beyond asimov: the three laws of responsible robotics. In: IEEE intelligent systems, Issue 14–20 July/Aug 2009
21. Sullins JP (2012) Robots, love and sex: the ethics of building a love machine. In: IEEE transactions on affective computing, vol 3, no 4, pp 389–409
22. Levy D (2008) Love and sex with robots: the evolution of human-robot relationship. Harper Perennial, London
23. Danaher J. Philosophical disquisitions: the ethics of robot sex. IEET: Institute for Ethics and Emerging Technologies. http://ieet.org/index.php/IEET/more/danaher20131014#When:11:03:00Z
24. McArthur N. Sex with robots, the moral and legal implications. http://news.umanitoba.ca/sex-with-robots-the-moral-and-legal-implications
25. Weltner A (2011) Do the robot. New Times 25(30), 23 Feb 2011. http://www.newtimesslo.com/news/5698/do-the-robot/
26. Brown (2013) HCRI: Humanity-centered-robotics-initiative, raunchy robotics—the ethics of sexbots, 18 June 2013. http://hcri.brown.edu/2013/06/18/raunchy-robotics-the-ethics-of-sexbots/
27. http://anneofcarversville.com/storage/2210realdoll1.png
28. http://s2.hubimg.com/u/8262197_f520.jpg

Chapter 12
Mental Robots

Yes, it's our emotions and imperfections that make us human.
Clyde Dsouza

The ability to improve behavior through learning is the hallmark of intelligence and thus the ultimate challenge of AI and robotics.
Maja J. Mataric

12.1 Introduction

Modern robots are designed to possess a variety of capabilities where perception, processing, and action are embodied in a recognizable human-like or animal-like form in order to emulate some subset of the physical, cognitive, intelligent and social dimensions of the human (and animal) behavior and experience. Human-like and animal-like robotics attracts electrical, mechanical, computer, and control engineers, but also philosophers, psychologists, neurobiologists, sociologists, and artists, all over the world, to contribute. Robotics with this kind of properties aim to create robot beings able to interact with humans beings, rather than replacing them, to do more sociable work. The ethical implications of this type of robots, which are called '*mental robot's*', were extensively studied in the previous chapters. In the present chapter we will be concerned with the basic issues of the mental (brain-like) features of them. Clearly, in all types of robots the mechanical and control part of them requires a sufficient degree of physical embodiment. But the mental part which involves the brain-like capabilities needs the development of robot-world environment, and embodied cognition and action. The brain-like capabilities need the development of robot-world environment, and embodied cognition and action. The capability of a mental robot to adapt to, learn from, and develop with its surrounding world (which represents the robot's interaction with its world) is tightly related to whether the robot will survive in this world. Mental robots represent a kind of *artificial—life* systems which according to Langton exhibit behaviors

S.G. Tzafestas, *Roboethics*, Intelligent Systems, Control and Automation: Science and Engineering 79, DOI 10.1007/978-3-319-21714-7_12

characteristic of natural living systems, that are created by synthesizing life-like behaviors within computer and other artificial media.

The purpose of this chapter is:

- To introduce the reader to the five principal mental abilities of modern robots, namely: cognition intelligence, autonomy, consciousness, and conscience (at a conceptual and fundamental philosophical level).
- To provide a brief discussion of the specialized cognitive/intelligence capabilities of learning and attention of these robots.

The material of this chapter is provided as a complement to the chapters on *artificial intelligence* (Chap. 3) and the *world of robots* (Chap. 4) in order to give the reader a better spherical picture of the robots that require an ethical consideration in both their design and use in our society.

12.2 General Structure of Mental Robots

Today much effort and money is given to create cognitive, intelligent, autonomous, conscious, and ethical robots (robots with conscience), for servicing the human beings in several ways as described in previous chapters (healthcare, assistance of low mobility people, mentally impaired people, etc.).

In general, sociable robots (anthropomorphic, zoomorphic) that aim to interact with people in human-like ways involve two major parts:

- *Physical or body part* (mechanical structure, kinematics, dynamics, control, head, face, arms/hands, legs, wheels, wings, etc.).
- *Mental or thinking part* (cognition, intelligence, autonomy, consciousness, conscience/ethics, and related processes, such as learning, emotions, etc.).

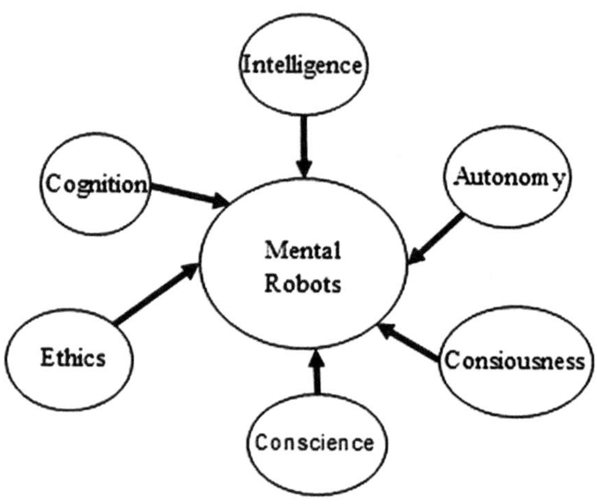

Fig. 12.1 The five constituent elements of the mental part of modern robots

Figure 12.1 illustrates the contribution of the areas of cognition, intelligence, autonomy, consciousness, conscience and ethics in achieving mental robots.

These elements are overlapping and interrelated in complex ways which are not yet uniquely and completely defined and described psychologically and philosophically. The actual design and implementation of mental robots is now at a good level of advancement, but much has to be done to achieve a truly mental robot possessing all human mental abilities. Making robots that act mentally like humans is never an easy job.

12.3 Capabilities of Mental Robots

In the following we give a brief outline of the mental capabilities required for having a mental robot.[1]

12.3.1 Cognition

Cognition refers to the full functioning of the brain at the higher level, not directly involving the details of the neurophysiological brain anatomy. It is not a distinct module of the brain, or a component of the mind that deals with rational planning and reasoning or acts on the representation acquired by the perception apparatus. Cognition is studied within the frameworks of psychology and philosophy [1].

Cognitive robotics is an important emerging field of robotics that cannot be defined in a unique and globally accepted way. The philosophical aspects of cognition can be considered from two points of view, i.e., (i) philosophy in cognitive science, and (ii) philosophy of cognitive science. The first deals with the philosophy of mind, philosophy of language, and philosophy of logic. The second deals with questions about cognitive models, explanations of cognition, correlations of causal and structural nature, computational issues of cognition, etc. Cognitive robotics is the engineering field of embodied cognitive science, and comes from the field of cybernetics initiated by Norbert Wiener as the study of communication and control in living organisms, machines and organizations [2].

Among the many different approaches for studying and building cognitive robots the following three approaches are most popular and offer good generic paradigms [3].

[1]The mechanical capabilities of robots (medical, assistive, social, etc) are studied separately by robot mechanics and control.

- **Cognivist approach** which is based on symbolic information representation and processing. Cognivism is closely related to Newell's and Simon's 'physical symbol system' approach to artificial intelligence [4].
- **Emergent systems approach** which embraces connectionist structures, dynamic structures, and enactive systems [3]. These systems are based on the philosophical view that cognition is an emergent, dynamic and self-organizing process, in contrast to cognivist approaches that see cognition as symbolic, rational, encapsulated, structured and algorithmic [5]. *Connectionist systems* are massive parallel processing systems of no symbolic distributed activation patterns, and involve the neural network or neural computing systems [6]. *Dynamic models* describe the perception action cognitive processes and self-organize into metastable patterns of behavior [7]. *Enactive systems* are based on the philosophical view that cognition is a process by which the aspects that are required for the continuous existence of a cognitive agent are brought out or enacted, and co-determined by the agent during its interaction with its environment. This means that nothing is given a priori and so there is no need for symbolic representations. Enactive systems are self-produced (autopoietic), i.e., they merge as coherent systemic entities.
- **Hybrid cognivist-emergent-systems approach** that combines the best features of the previous two approaches.

The basic requirements for the development of cognitive robots include:

- *Embodiment* (physical sensory and motor interfaces compatible with the model of cognition).
- *Perception* (attention on the goal of action, perception of objects, and ability to learn hierarchical representations).
- *Actions* (early movements with small number of degrees of freedom and navigation based on dynamic ego-centric path integration).
- *Adaptation* (transient and generalized episodic memories of past experiences).
- *Motivation* (explorative motives).
- *Autonomy* (presentation of homeostasis processes, behaviors for exploration and survival).

Figure 12.2 gives a pictorial illustration of the processes involved in the human cognition cycle where the dotted "Consolidation" link indicates that it occurs outside the cognitive cycle.

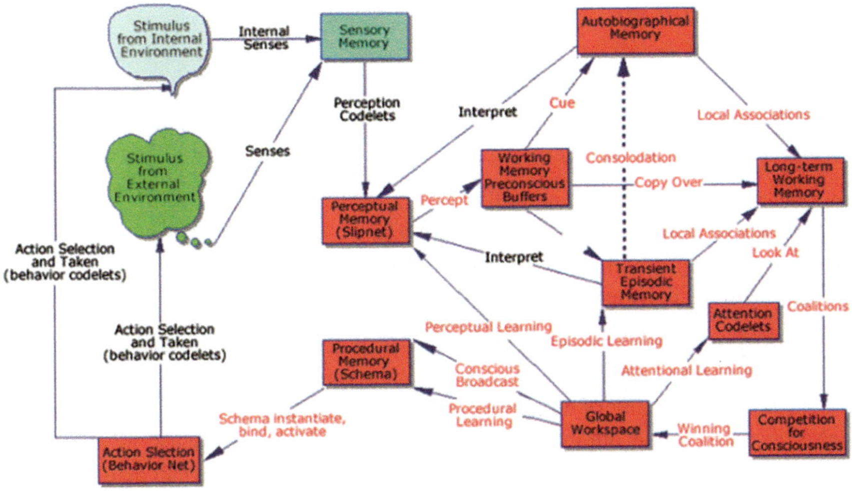

Fig. 12.2 Human cognition cycle. *Source* http://www.brains-minds-media.org/archive/150/RedaktionBRAIN1120462504.52-3.png

12.3.2 Intelligence

Intelligence is the general human cognition ability for solving problems in life and society. Individuals differ from one another in their ability to understand complicated processes, to adapt effectively to the environment, to learn from experience, and to reason under various conditions. These differences may be substantial and are not consistent since they actually vary from situation to situation or from time to time. Psychology has developed several methods for measuring or judging the level of a person's intelligence; the so-called psychometric intelligence tests. On the basis of them developmental psychologists study the way children come to think intelligently, and distinguish mentally retarded children from those with behavior problems.

With the advancement of the computer science field, many attempts were initiated to create and study machines that posses some kind and level of intelligence (problem solving ability, etc) analogous to human intelligence, which as we have seen in Chap. 3 is still the subject of strong controversy. Robots represent a class of machines that can be equipped with some machine intelligence which can be tested using the 'Turning Test'. As with human intelligence, in robot and machine intelligence many philosophical questions came to be addressed, the two dominant of which are whether intelligence can be reproduced artificially or not, and what are the differences, between human intelligence and machine intelligence.

12.3.3 Autonomy

Autonomy is a concept that has a multiplicity of meanings and ways in which it can be understood, interpreted and used in human's life. Four closely related fundamental meanings are the following:

- The capacity to govern oneself.
- The actual condition of governing oneself.
- The 'sovereign' authority to govern oneself.
- An ideal of character.

The first meaning is understood as the basic autonomy (minimal capacity) which refers to the ability to act independently, authoritatively, and responsibly.

The second meaning of autonomy mirrors the one's entitlement to certain liberal rights that determine our political status and freedoms. However, in actual life, having the capacity to govern ourselves does not imply that we can actually do so.

The third meaning distinguishes *dejure* and *defacto* autonomy, where the former refers to the moral and legal right to self-government, and the latter to the competence and opportunities required for exerting that right.

Finally, the fourth meaning (autonomy as ideal) refers to our moral autonomous agency predicated upon 'autonomy virtues' with the aid of which we can correctly guide our agency and orientate public social policies concerned with fostering autonomy.

In robotics, autonomy is interpreted as independence of control. This meaning implies that autonomy characterizes the relation between the human designer and controller, and the robot. A robot's degree of autonomy is increased if the robot possesses increasing abilities of self-sufficiency, situadness, learning, development, and evolution. The two forms of robot autonomy are [8]:

- *Weak autonomy* (the robot can operate and function free from any outside intervention).
- *Strong (or full) autonomy* (the robot can make choices on its own and is authorized to activate them).

If a robot combines intelligence with autonomy, then it said to be an 'intelligent autonomous robot'.

12.3.4 Consciousness and Conscience

These two aspects of human performance have been the subject of extensive and deep studies made by psychologists and philosophers, in their attempt to explain the functions of the human brain:

- *Consciousness* refers to the issue of how is it possible to observe some of the processes that take place in our brain.

- *Conscience* refers to the issue of how is it possible to acquire and use the knowledge of what is right or wrong.

In philosophy, consciousness is interpreted as the mind or the mental abilities exhibited by thoughts, feelings and volition. In psychology, consciousness has several meanings, e.g., awareness of something for what it is or the thoughts and feelings, collectively, of an individual (or a society).

Extending consciousness to machines and robots is not an easy issue. According to Haikonen [9], for a robot to be conscious it is required to have some kind of mind, to be self-motivated, to understand emotions and language and use it for natural communication, to be able to react in emotional way, to be self-aware, and to perceive its mental content as immaterial. In [13] an engineering approach that would lead towards cognitive and conscious machines is outlined, using neuron models and associative neural networks made from the so-called 'Haikonen associative neurons'.

In [10], Pitrat provides a comprehensive study of human consciousness and conscience and investigates whether it is possible to create 'artificial beings' (robot beings, etc.), that possess some capabilities analogous to those that consciousness and conscience give to human beings. He argues that 'if a system or machine (like a robot) generates a behavior similar to ours, it is possible that we are also using it'. However, as he says, many artificial intelligence workers are not interested to understand how we work, but to realize systems that are as efficient as possible, with necessarily modeling of the brain.

Human beings have a warning mechanism built into them which warns them that they are malfunctioning. In other words, the resulting warning signals manifest that a human does not behave in harmony with his own values and beliefs. This mechanism is the *human conscience.* Building a mechanism of this type we get a robot with conscience which is determined by the values and norms embedded in it (*top down robot conscience* approach). The question whether it is possible to build conscience (and ethics) into robots or not, has been a diachronic issue of study in psychology, philosophy and robotics.

12.4 Learning and Attention

Two other mental capabilities of modern robots that belong to cognition and intelligence are the following:

- Learning
- Attention

12.4.1 Learning

Learning is one of the most fundamental cognition and intelligence processing abilities which involves acquisition of new behavior and knowledge, and occurs continuously thought a person's life. The two basic learning theories are *behaviorism* (i.e., particular forms of behavior reinforced by the teacher to shape or control what is learned), and *cognitive theory,* which unlike behaviorism focuses on what is going inside the learner's mind. In other words, cognitive theorists advocate that learning is not just a change in behavior, but a change in the way a learner thinks, understands or feels. The two major branches of cognitive theory are:

- *Information processing model* (The student's brain has an internal structure which chooses, retrieves, stores and processes incoming information).
- *Social interaction model* (People learn by watching other people performing a behavior and the outcome of this behavior). This model was coined by Bandura [11].

Fundamental necessary conditions for effective social learning are: *attention* (one pays attention to learn), *retention* (remembering what one pays attention to), *reproduction* (reproducing the image), and *motivation* (having a good reason to imitate).

In humans, learning is performed according to the following styles:

Visual (learn through seeing), *auditory* (learn through listening), and *kinesthetic* (learn through moving, doing, and touching).

Robot learning can be performed by using methods and processes similar to human learning [12]. Robots can be taught concepts, how to acquire information, how to use their own sensors, to express emotions, to navigate, and even to teach-themselves. Social robots can learn via neural networks, via reinforcement, via evolutionary learning, and through imitative learning (learning by imitation) which has been extensively studied in the 'developmental psychology field' [13].

Critical questions that have to be answered for achieving such 'developing' social robots include [14].

- How a social robot can direct its own development?
- How this development is motivated and guided?
- What bounds (if any) should be imposed?

These questions may never have a definite answer, but a balance of human input, self-development, and real-world interaction seems to be feasible and has actually been realized in existing humanoid (anthropomorphic) social robots.

Figure 12.3 gives a detailed pictorial illustration of the human learning styles.

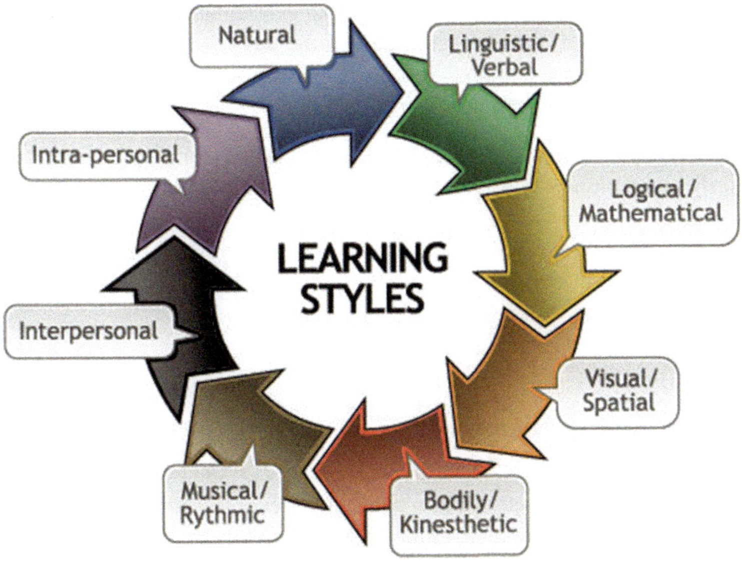

Fig. 12.3 Detailed human learning styles. *Source* http://4.bp.blogspot.com/-z4yYFo3zPtQ/UFjOvIP32sI/AAAAAAAAACk/fkqo5oqXMyY/s1600/learning-styles2.jpg

12.4.2 Attention

Attention is the human cognitive ability to focus selectively on a specific stimulus, sustaining that focus and shifting it at will, i.e., the ability to 'concentrate'. It is a concept of cognitive psychology which refers to how humans actively process specific information available in the environment. Attention is important to learning, since learning is optimally efficient when a person is paying attention. The attention ability of selecting the potentially relevant parts out of a large amount of sensory data enables interactions with other human beings by sharing attention with each other. This ability is of great importance in robotics, where the computational modeling of human attention is a key issue in human-robot interaction. The ability of a robot to detect what a human partner is attending to, and to act in a similar manner enables intuitive communication which is an important desired skill for a robotic system.

Attention is triggered by two different types of influences [15]:

- *Stimulus-driven* (i.e., affected by bottom-up influences).
- *Goal-directed* (i.e., fueled by top-down influences).

Attention is the process by which a being allocates perceptual resources to analyze an area of the surrounding world in detriment of others. This allocation can be done in two ways:

- Explicit reorientation of sensors (e.g., head reorientation in the case of visual and auditory sensors) to a specific part of the world (*overt attention*).
- Deployment of computational resources for processing a specific part of the sensory information stream (*covert attention*)

Overt attention is the direct cause of active perception [16]. It can be either voluntarily or involuntarily driven by automatic orientation of sensors. Usually, the overall attention starts with overt attention which is followed by covert_mechanisms. Basically, the involuntary attention process is stimulus-driven, but it is also modulated by goal-directed influences through *attention sets* which impose as a 'priority measure' the task relevance.

12.5 Concluding Remarks

In this chapter we have discussed at a conceptual level the human brain capabilities which robot cists are attempting to embody in modern service and social robots at one or the other degree. These include the five primary features of cognition, intelligence, autonomy, consciousness and conscience, and two special capabilities involved in cognition and intelligence, namely learning and attention. From a philosophical viewpoint our discussion was restricted to the *ontological* and *epistemological* issues of these capabilities. *Philosophy* (from the Greek word *φιλοσοφία* = *philosophia* = love of wisdom) involves the following principal subfields [17, 18]: *Metaphysics* (*ontology*) which studies the concept of '*being*' (i.e., what we mean when we say that something '*is*'), *epistemology* that studies issues related to the nature of knowledge (e.g., questions such as: 'what can we know?', 'how do we know anything?, and 'what is truth?', *teleology* that asks about the aims and purposes of 'what we do' and 'why we exist', *ethics* that studies the 'good and bad' and 'right and wrong', *aesthetics* that studies the concepts of beauty, pleasure and expression (in life and art), and *logic* which studies the issue of reasoning, including questions like 'what is rationality?', 'can logic be computationally automated?', etc.

For anything we care to be interested we have a philosophy which deals with the investigation of its fundamental assumptions, questions, methods, and goals, i.e., for any X there is a philosophy which is concerned with the ontological, epistemological, teleological, ethical and aesthetic issues of X. Thus, we have philosophy of science, philosophy of technology, philosophy of biology, philosophy of computer science, philosophy of robotics (robophilosophy), etc. *Ontology* is classified in several ways, e.g., according to its truth or falsity, according to its potency, energy (movement) or finished presence, and according to the level of abstraction (upper ontologies, domain ontologies, interface ontologies, process ontologies). *Epistemology* involves two traditional approaches [19, 20]: *rationalism* according to which knowledge is gained via reasoning, and *empiricism* according to which knowledge is acquired through sensory observation and measurements. Philosophers agree that both these approaches to knowledge are required, and that to a certain extent they complement and correct each other. Comprehensive studies of philosophical aspects of artificial intelligence and mental robots, focusing on roboethics and sociable robots, are presented in [21–26].

References

1. Stenberg R (1991) The Nature of Cognition. MIT Press, Cambridge, MA
2. Wiener N (1948) Cybernetics: control and communication in the animal and the machine. MIT Press, Cambridge, MA
3. Vernon D, Metta G, Sandini G (2007) A survey of artificial cognitive systems: implications for the autonomous development of mental capabilities in computational agents. IEEE Trans Evol Comput 1(2):151–157
4. Newell A (1990) Unified theories of cognition. Harvard University Press, Cambridge, MA
5. Vernon D (2006) Cognitive vision: the case for embodied perception. Image Vision Comput 1–14
6. Arbib MA (ed) (2002) The handbook of brain theory and neural networks. MIT Press, Cambridge, MA
7. Thelen E, Smith LB (1994) A Dynamic system approach to the development of cognition and action, bradford book series in cognitive psychology. MIT Press, Cambridge, MA
8. Beer JM, Fisk AD, Rogers WA (2012) Toward a psychological framework for levels of robot autonomy in human-robot interaction, Technical Report HFA-TR-1204, Schoolof Psychology, Georgia Tech.
9. Haikonen PO (2007) Robot brains: circuits and systems for conscious machines. Wiley, New York
10. Pitrat J (2007) Artificial beings: the conscience of conscious machine. Wiley, Hoboken, NJ
11. Bandura A (1997) Social learning theory. General Learning Press, New York
12. Alpaydin E (1997) Introduction to machine learning. McGraw-Hill, New York
13. Breazeal C, Scasselati B (2002) Robots that imitate humans. Trends Cogn Sci 6:481–487
14. Swinson ML, Bruener D (2000) Expanding frontiers of humanoid robots. IEEE Intell Syst Their Appl 15(4):12–17
15. Corbetta M, Schulman GL (2002) Control of goal-directed and stimulus-driven attention in the brain. Nature Rev Neurosci 3:201–215
16. Aloimonos J, Weiss I, Bandyopadhyay A (1987) Active vision. Int J Comput Vision 1:333–356
17. Stroud B (2000) Meaning, understanding, and practice: philosophical essays. Oxford University Press, Oxford
18. Munitt MK (1981) Contemporary analytic philosophy. Prentice Hall, Upper Saddle River, NJ
19. Dancy J (1992) An introduction to contemporary epistemology. Wiley, New York
20. BonJour L (2002) Epistemology: classic problems of contemporary responses. Rowman and Littlefield, Lanthan, MD
21. Boden MA (ed) (1990) The philosophy of artificial intelligence. Oxford University Press, Oxford, UK
22. Copeland J (1993) Artificial intelligence: a philosophical introduction. Wiley, London, UK
23. Catter M (2007) Minds and computers: an introduction to the philosophy of artificial intelligence. Edinburgh University Press, Edinburgh, UK
24. Moravec H (2000) Robot: mere machine to trancedent mind. Oxford University Press, Oxford, UK
25. Gunkel DJ (2012) The machine question: critical perspectives on ai, robots, and ethics. MIT Press, Cambridge, MA
26. Seibt J, Hakli R, Norskov M (eds) (2014) Sociable robots and the future of social relations. In: Proceedings of Robo-Philosophy, vol 273, Frontiers in AI and Applications Series. Aarchus University, Denmark, IOS Press, Amsterdam, 20–23 Aug 2014

Index

S.G. Tzafestas, *Roboethics*, Intelligent Systems, Control and Automation: Science and Engineering 79, DOI 10.1007/978-3-319-21714-7

9783319217130